APEX 8

The Art of Persuasion in a Technical World

Daren L. Fields

To LeTricia,

Who believed in this book before a single word was written.
Your unwavering support, encouragement, and faith in my abilities
have been the foundation upon which these pages stand.
You pushed me to strive for more
when I thought I had reached my limits,
and never once questioned my ability
to share these ideas with the world.
This book exists because you exist in my life.

To my children,

Thank you for the many sacrifices you made while your father was
running around the country or traveling around the world doing
what he loves to do. Hopefully some of you will continue to—or may
even begin to—follow in my footsteps and begin the joy of persuad-
ing people to do the things they really should be doing.

You have always been my motivation to work hard on my craft
and to understand everything I could in order to be successful,
and for that I am forever grateful to you.

CONTENTS

CHAPTER 1

THE COMPLETE TRANSFORMATION
OF TECHNICAL SELLING

"I appreciate you all coming in today, but there's going to be a change in the agenda."

The procurement director's expression was a mix of apology and efficiency as she motioned to the monitor in the room. As you looked at the screen, you saw a pulsating circle—the CTO's pet project, an AI assistant they've named CORBIN (Cognitive Optimization for Revenue, Behavior, Intelligence & Navigation).

"CORBIN has been analyzing all the solutions we've seen this week," she explained, "and has identified where the core functionalities overlap. We'd like to focus exclusively on your differentiators today."

The screen shifted to display a matrix comparing feature sets across competitors, with capability gaps highlighted in red and your presumed advantages in green.

"As you can see," the director continued, "we're already familiar with the standard capabilities of your platform. What we need to understand is how your implementation team will collaborate with our software developers and AI engineers during deployment. We're particularly concerned about your security protocols at all integration points and how you'll prevent data leakage when our systems communicate. Additionally, how will you handle the inevitable cultural resistance from teams who fear disruption?"

The carefully crafted presentation you'd spent weeks preparing was suddenly obsolete before you'd even begun. Your detailed walkthrough of features would now waste precious minutes in a meeting that had already pivoted to complex human and security concerns not covered in your standard deck.

Welcome to technical selling in the AI era—where artificial intelligence hasn't just changed the game; it's created an entirely new playing field.

The transformation we're experiencing isn't incremental—it's existential. In the span of just a couple of years, artificial intelligence has rendered decades of traditional sales approaches obsolete. Companies that have embraced this shift are seeing dramatic results: sales teams using AI achieve 26% higher revenue growth rates than their non-AI counterparts[1], and deals where AI-recommended actions are followed see win rates increase by an impressive 50%[2]. Meanwhile, those clinging to pre-AI methods are struggling in an environment where average win rates have declined from 23% to 19% in just two years[3].

What makes this revolution so profound isn't just the technology itself, but how completely it has inverted the power dynamics of selling:

- **Information asymmetry has reversed:** The knowledge advantage that once belonged to sellers has shifted to AI-assisted buyers. Your prospects' AI assistants now know more about your offerings, pricing, and implementation history than your average sales rep. They've analyzed every case study, forum post, and customer review before you've even opened your mouth.

- **Decision intelligence has democratized:** Mid-level stakeholders equipped with AI support now wield analytical capabilities once reserved for the C-suite. A procurement manager with the right AI tools can run sophisticated TCO analyses, risk projections, and implementation simulations that previously required teams of consultants.

- **Performance transparency is inescapable:** Every claim you make is instantly verifiable. Every commitment you offer is recorded, transcribed, and compared against your organization's historical delivery metrics. AI-enabled customers track promise fulfillment with a precision that makes accountability non-negotiable.

- **Value articulation has been quantified:** Generic value propositions die on contact with today's AI-equipped buyers. They demand mathematically precise ROI projections tailored to their specific operational context—and they have the tools to validate your calculations in real-time.

This isn't simply about working with new tools—it's about surviving in an environment where the fundamental nature of business decision-making has been rewired by artificial intelligence. Organizations with effective sales enablement strategies that leverage AI are achieving 49% higher win rates on forecasted deals[4]. The gap between AI-fluent sellers and traditional representatives grows wider every quarter, with sales professionals increasingly turning to AI to automate manual processes and focus on higher-value activities[5].

Why Listen to Me?

Before diving into the philosophy behind Apex 8, it's fair to ask: why take my word for it?

I've spent more than two decades in high-stakes technical sales, leading teams and building strategies across complex buying environments. I've worked on everything from enterprise software to advanced manufacturing systems, and I've coached dozens of teams through deals that stretched over quarters and spanned continents.

For instance, in a recent ERP implementation, applying Apex 8 principles helped us uncover a critical requirement that never appeared in the formal documentation. During an informal dinner, an executive mentioned their urgent need to integrate AI technology into their distributor sales program. We quickly pivoted to build a prototype of an intelligent chatbot that accessed real-time ERP data. Without this discovery—made possible by looking beyond standard requirements—we would have lost the deal entirely.

What I'm sharing in this book isn't theory. It's a distillation of what works when the stakes are high, the problems are complex, and the buyer is better informed than ever because of AI. To navigate this new reality effectively, you need a framework and a methodology that acknowledges how learning actually happens in both you and your customer. This learning element isn't just in sales, but it becomes quite evident when it is. That's where the concept of conscious competence becomes essential to your survival—not just as a learning theory, but as your practical advantage in AI-enhanced selling.

The Learning Ladder

The concept of conscious competence comes from adult learning theory. It describes four levels of skill acquisition:

CONSCIOUS COMPETENCE LEARNING MODEL	
Unconscious Incompetence You don't know that you don't know	**Conscious Incompetence** You know that You don't know
Conscious Competence You know that You know	**Unconscious Competence** You don't know that You know

1. **Unconscious incompetence:** You don't know what you don't know.
2. **Conscious incompetence:** You realize your limitations and start seeking improvement.
3. **Conscious competence:** You can perform well with effort and attention.
4. **Unconscious competence:** The skill becomes second nature.

But in today's environment, relying solely on unconscious competence can actually limit your effectiveness. It makes it harder to adapt in unfamiliar situations, coach others clearly, or collaborate with AI systems that require structured input. Just as you need to articulate your thinking to peers or leaders, you'll increasingly need to explain it to your AI tools. If you can't describe what you're trying to achieve—whether to a carbon-based colleague or a silicon-based assistant—you'll struggle. That's why conscious competence has become not just a learning phase, but a communication skill. Learn to speak your process with precision when needed, then let unconscious flow take over in the customer moment.

Meta Moments:
Your Human Advantage in an AI-Dominated World

While AI can process information faster than any human, it can't yet create true "meta moments"—those powerful instances where you fundamentally reshape how a buyer sees their entire problem.

So what exactly is a meta moment in today's AI-enhanced selling environment?

It's when you step beyond the transaction to help customers see possibilities they hadn't considered. In a world where AI handles the routine parts of selling, these meta moments become your most valuable contribution. It's like the difference between GPS navigation and actually being a local guide—AI can provide directions, but only you can show the shortcuts, point out the hidden gems, and explain why the locals avoid certain routes.

Think about your last breakthrough sales conversation. There was likely a moment when you saw the customer's perspective shift— their eyes widened, they leaned forward, they started asking different questions. That's a meta moment, and it's something AI still struggles to create.

As you grow in your practice, these meta moments begin to surface:

- **You start breaking down sales into micro-moments.** AI tools help you spot subtle shifts in tone, timing, and traction—allowing you to respond with precision before anyone else even notices.

- **You realize that truth, delivered with speed and clarity, is your greatest sales tool.** In a world where AI can slice through noise, your job is to get to the heart of the matter faster—and help your audience do the same.

- **You notice patterns across seemingly unrelated customer needs**—revealing deeper truths about what's really driving demand.

- **You realize you're not just selling a product; you're redefining a category**—positioning your solution in a new strategic light.

- **You begin to shape how your customers think about their challenges**—not just responding to problems but reframing them entirely.

- **You see how the way you sell starts to establish a new industry benchmark**—a standard others begin to follow.

- **You find yourself creating a new language or framework for how others understand the problem space**—and influencing how the entire conversation evolves.

- **You stop selling altogether—and start influencing how others decide**.

As AI accelerates logic, clarity, and trust, your role shifts from persuader to trusted navigator. Precision replaces pressure.

Operating at this elevated level of salesmanship also carries a deeper responsibility. With the power to influence high-stakes decisions comes the imperative to do so ethically and transparently.

The Ethical Foundation

Truth is your competitive advantage.

I remember a deal for a product lifecycle management system in the automotive sector that faced a moment of real risk. We were white-labeling the platform—we didn't own it—but we understood it better than the publisher. Why? Because we had deeper knowledge of auto manufacturing. Our team had significant experience, and we had a rock-solid implementation plan.

One of the decision-makers asked directly about our relationship with the software vendor. I didn't spin it. I explained exactly how it worked—who owned what, how we got paid, and why our team would still be the ones delivering results.

They could have walked away. But they didn't. They stayed because they trusted us to solve their problem. They realized that saving a few points by working directly with the publisher wasn't worth losing the people who actually knew how to make it work.

That moment of honesty didn't hurt us—it locked us in. It was a $1.2 million annual subscription deal.

The Laws of Exchange: Transformed by AI

1. The Law of Transparency
AI has surfaced previously hidden information

2. The Law of Reciprocity
AI has reduced friction in value exchange

3. The Law of Earned Trust
AI has scrutinized authenticity

4. The Law of Compounding Value
AI has accelerated time-to-value

The fundamental principles that govern value exchange in sales have existed for centuries, but AI has radically transformed how they operate in today's selling environment:

1. **The Law of Transparency:** In the pre-AI era, sellers could control the flow of information. Today, AI instantly reveals pricing discrepancies, implementation challenges, and competitor comparisons. What was once hidden is now exposed with a simple query. This means transparency isn't just ethical—it's your only viable strategy when buyers' AI assistants can fact-check your claims in real time.

2. **The Law of Reciprocity:** AI has supercharged the share economy by reducing friction in value exchange. When you

freely provide insights through content, tools, or consultations without demanding immediate returns, AI amplifies this generosity—distributing your expertise to exponentially more potential buyers. This creates a powerful motivation for others to reciprocate when they receive genuine value.

3. **The Law of Earned Trust:** As buyers become increasingly skeptical of manufactured authenticity, AI systems are being trained to detect inconsistencies in seller claims and behavior. The new currency in sales isn't just delivering value—it's building trust through consistently ethical behavior that withstands AI-powered scrutiny.

4. **The Law of Compounding Value:** AI dramatically accelerates how quickly buyers can implement and see results from your solution. This compression of time-to-value means buyers expect faster returns and clearer measurements. Your ability to articulate and accelerate this value velocity becomes crucial.

Mastering these AI-transformed laws not only fosters ethical selling—it provides a strategic advantage in a world where artificial intelligence continually reshapes the balance of power between buyers and sellers.

Apex 8 Framework: A Preview

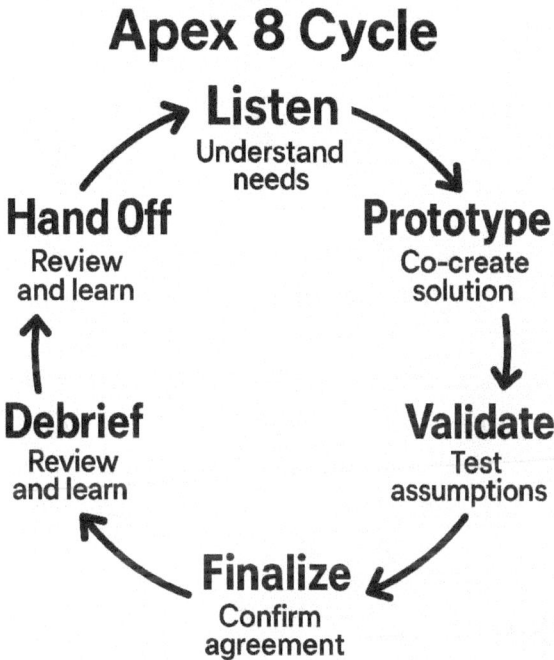

Apex 8 Cycle

Listen — Understand needs
Prototype — Co-create solution
Validate — Test assumptions
Finalize — Confirm agreement
Debrief — Review and learn
Hand Off — Review and learn

Here are the eight steps at the core of this methodology:

1. **Discover:** Understand the buyer's reality and what's missing from their current approach.
2. **Prototype:** Co-create a vision of the future with tangible, visual tools.
3. **Validate:** Build confidence through verification, testing, and stakeholder input.
4. **Rehearse:** Prepare like a performer—not just to present, but to connect.

5. **Present:** Deliver a message that resonates intellectually, emotionally, and politically.
6. **Finalize:** Help buyers move past fear into confident commitment.
7. **Debrief:** Learn from every opportunity, win or lose.
8. **Handoff:** Ensure value is realized and momentum continues post-sale.

Each step in Apex 8 is designed to reinforce the foundations we've established in this chapter. Discovery sharpens awareness. Prototype invites co-creation. Validation builds trust through shared truth. Rehearsal transforms insight into performance. All of it is grounded in conscious competence, ethical execution, and the laws of exchange that govern how value flows. This isn't just a sequence—it's a system designed for integrity, adaptability, and mastery in the AI-powered era of selling.

Your First 30 Days in AI-Powered Technical Selling

The gap between sellers who master AI and those who merely use it is already widening. Here's your action plan for the first month to ensure you're on the right side of that divide:

1. Create Your AI Sales Intelligence System

Today's AI can turn every customer interaction into actionable intelligence that compounds over time. Start by selecting an AI-powered conversation platform (Otter.ai, Fireflies.ai, or Avoma) to automatically record, transcribe, and analyze all your sales conversations. Configure it to detect buying signals, objection patterns, and competitive mentions. Within two weeks, you'll have a searchable database of insights that would have taken months to compile manually. Pro tip: Ask your AI to identify the questions you're consistently failing to answer effectively—these gaps represent your fastest path to improvement.

2. Build Your Technical Sales Co-Pilot

Subscribe to a professional-grade large language model (Claude Pro, ChatGPT Plus, or industry-specific alternatives) and begin training it on your specific sales context. Upload your product documentation, past winning proposals, competitor analyses, and industry reports. Create custom prompts for pre-call research, objection handling, and simplifying technical explanations. By personalizing your AI co-pilot, you're developing a force multiplier that adapts to your unique selling environment. Experiment with having it draft personalized follow-ups based on call transcripts—customers won't believe how relevant your communications suddenly become.

3. Map Your Competency Gaps with AI Analysis

Use your AI tools to analyze where you fall on the competency ladder for each aspect of technical selling. Have your AI assistant review your last five calls to identify patterns in your questioning technique, value articulation, and technical explanation clarity. Be brutally honest about what the data reveals—are you unconsciously incompetent in certain critical areas? This baseline assessment becomes your personalized development roadmap.

4. Create a Meta Moment Trigger System

Program your AI assistant to alert you to potential meta moment opportunities during customer conversations. These might include detecting when a customer expresses frustration with existing approaches, mentions failed initiatives, or struggles to articulate their vision. By recognizing these openings in real time, you can pivot from routine discussions to transformative insights. Many sellers report that just three well-timed meta moments can completely reframe a complex sale.

5. Join the AI Sales Collective

Find a community of technical sellers who are also embracing AI transformation. Whether through industry groups, online forums, or dedicated Slack channels, connect with peers who will challenge your thinking and share emerging best practices. The velocity of AI advancement means isolated sellers fall behind rapidly—collective intelligence becomes your competitive advantage. Share what you're learning about AI implementation while absorbing others' insights, creating a powerful feedback loop of continuous improvement.

The beauty of this AI era is that your improvement rate is now limited only by your commitment to learning, not by the availability of feedback or resources. With your own private AI system analyzing your every move and suggesting improvements, you're free to experiment, fail, and grow without judgment. You'll be astonished at how quickly you can advance when you leverage AI not just as a tool, but as a learning accelerator that never sleeps.

In the next chapter, we'll explore the first step of Apex 8: Discovery. We'll examine how this critical phase has been transformed in the AI era and how you can leverage both human insight and technological tools to build a deeper understanding of your buyer's reality.

But remember—the philosophical foundation we've established in this chapter underlies everything that follows. Conscious competence, ethical selling, and the laws of exchange aren't just theoretical concepts; they're practical principles that will guide your application of each step in the Apex 8 methodology.

The technical seller who masters these principles while developing expertise in the eight steps of Apex 8 will not be replaced by AI; they will be empowered by it, elevated to a higher level of performance than what was possible in previous eras of professional selling.

The future belongs to the conscious, competent technical seller. Let's build that future together

[1] Salesforce State of Sales Report, 2024
[2] Gong Labs Research, 2024
[3] Sales Industry Trends Report, 2024
[4] G2 Sales Enablement Statistics, 2025
[5] HubSpot State of AI Report, 2024

CHAPTER 2

DISCOVERY: UNCOVERING REALITY IN THE AGE OF AUGMENTED INTELLIGENCE

The New Game of Discovery

Sarah had thoroughly prepared for this discovery call. As a technical sales specialist for advanced manufacturing equipment with twelve years of experience, she recognized that effective discovery was the foundation of any successful deal. Her company, PrecisionTech, manufactured high-end automated assembly line systems for automotive components. For this meeting with Velocity Motors, a tier-one auto parts supplier, she had researched their production facilities, studied their product lines, analyzed their recent expansion announcements, and prepared her standard discovery questions regarding production volume and quality control issues.

However, twenty minutes into the plant tour, something felt amiss. The prospect—a Director of Manufacturing Operations—appeared disengaged, almost impatient.

"Let me save us both some time," he said finally, pausing beside a row of existing assembly robots. "I've already had four equipment vendors walk this floor this month. My AI procurement assistant analyzed your systems last night and generated a detailed comparison of your servo precision ratings, maintenance requirements, and integration capabilities against your competitors. What I really need to understand is how your assembly line would handle our specific challenge with variable material densities in our new composite parts, particularly the carbon-fiber-reinforced polymer components that keep failing QC at high production speeds."

Sarah sensed her carefully prepared presentation crumbling. The buyer already had more structured technical data about her equipment than she typically shared in first meetings. He had bypassed her planned qualification questions about basic needs and jumped straight to specialized engineering challenges she would normally address weeks later in the process.

Welcome to discovery in the AI era.

The first step in the Apex 8 methodology—Discovery—has transformed more profoundly than perhaps any other aspect of technical selling. The information asymmetry that once gave sellers an advantage has dramatically shifted. Today's buyers come armed with AI-gathered intelligence, automated analysis, and a technical understanding that would have been impossible just a few years ago.

But this doesn't diminish the importance of discovery; it elevates it. True discovery in technical selling has never been merely about gathering information. It's about uncovering reality—the buyer's current situation, their challenges, and their vision for the future. It's about making meaning from this information in ways that AI alone cannot.

In this chapter, we'll explore how to conduct discovery that goes beyond what AI can provide, creating insights that help buyers see their own situation with new clarity while positioning you as a trusted advisor rather than just another vendor in an AI-generated comparison matrix.

The AEO Revolution: How Buyers Research Now

To understand the new dynamics of discovery, we must first recognize how dramatically buyer research has changed. The era of traditional SEO (Search Engine Optimization) dominance—where buyers found solutions through carefully crafted Google searches—is rapidly fading. In its place has emerged AEO: Answer Engine Optimization.

This shift requires sellers to rethink their preparation strategy entirely. Today's buyers aren't just Googling; they're asking AI tools like Perplexity, ChatGPT, and Claude to generate comprehensive narratives about solutions, providers, risks, and implementation strategies before they ever speak to a human.

These tools don't return link lists; they produce authoritative-sounding answers by synthesizing content from blogs, case studies, technical documentation, and video transcripts. If you and your solution aren't appearing in these AI-generated narratives, you may be invisible to buyers, regardless of how perfect your fit might be.

For technical sellers, this means three critical adjustments:

First, you need to optimize for AI visibility, not just search ranking—creating clear, structured, educational content about your methodology and results that AI can easily find and reference.

Second, you must anticipate what information the buyer has already consumed. Before your call, they have likely asked something like, "Who are the top vendors solving X challenge in Y industry using Z technology?" You need to know what answer they probably received.

Third, you need to plan your discovery to go beyond what AI has already told them—focusing on contextual insights, organizational nuances, and challenges specific to their environment that no AI could adequately address.

When I talk about optimizing, anticipating, or planning for what buyers might have learned from AI, I'm not suggesting you do all this manually. Use your own LLM that you pay a subscription for! This is critical. When I say "you" should do these things, I mean the partnership between you and your AI assistant. For example, you can prompt your LLM with: "If a prospect did an AEO search on our product XYZ, what information would they likely receive?" or "Show me what the AI narrative probably looks like for our solution category."

This approach means you and your neural network, along with the LLM's neural network, can handle this preparation efficiently—without burning out your team or frustrating your customers with redundant questions. The best technical sellers are already using AI to prepare for AI-informed buyers.

This shift from SEO to AEO has profound implications for technical sellers:

1. **Buyers form opinions earlier:** Before your first interaction, buyers have often received AI-generated analyses that shape

their perception of your solution, your company, and even your industry.

2. **Technical depth arrives sooner:** Buyers can quickly access detailed technical information that previously would have required multiple sales conversations to reach.

3. **Comparison is constant:** Your solution is continually being evaluated against competitors in structured, side-by-side analyses generated by AI.

4. **Sentiment is quantified:** Subjective aspects like customer satisfaction and implementation experience are now quantified through AI analysis of reviews, case studies, and social mentions.

5. **Questions evolve faster:** As buyers get answers to their initial questions from AI, they quickly move to more sophisticated inquiries by the time they speak with you.

But this AI-powered research revolution also creates new opportunities for skilled technical sellers. The information AI provides is comprehensive but often lacks crucial context. It struggles with nuance, organization-specific variables, and the human factors that often determine success or failure in complex implementations.

This creates what I call the "AI Knowledge Gap"—the space between what AI can tell a buyer and what they truly need to understand to make an optimal decision. Identifying and filling this gap is the essence of modern discovery.

The AI Knowledge Gap

Despite their expanding capabilities and continuous improvement through training, today's AI systems have significant limitations when helping buyers research technical solutions. These gaps create rich opportunities for well-prepared sellers who understand how to identify and address them:

Context Blindness: AI excels at processing information but struggles with the organizational nuances that often determine success. It can't truly grasp your buyer's unique culture, their definition of winning, their traditions, or their collective wisdom. Your discovery must deliberately draw out these contextual elements that AI simply cannot see.

Experience Limitation: While AI aggregates others' experiences, it can't substitute for the hard-won knowledge gained from actually implementing solutions. It won't anticipate unpredictable human factors—like key stakeholders changing roles mid-project or competing priorities suddenly emerging—that experienced sellers have learned to plan for. Your value lies in recognizing these potential disruptions before they derail implementation.

Wisdom Deficit: Data processing isn't wisdom. AI lacks the pattern recognition that comes from years of seeing which problems truly matter versus which are merely distractions. It doesn't understand the value of timing—knowing when to push forward versus when to wait for organizational readiness. This visceral knowledge remains solidly human territory and a crucial part of your discovery approach.

Innovation Boundaries: AI generates but doesn't genuinely create. It works with existing information rather than envisioning entirely new possibilities. It won't suggest creative workarounds and novel approaches that stem from understanding the human elements of technical challenges. Your discovery must explicitly explore these innovation opportunities that AI overlooks.

Integration Blindness: AI often misses the subtle interdependencies between systems, processes, and people. It doesn't recognize when automation might create unexpected friction or when streamlining one department might impact another. Your discovery must actively search for these connection points—the relationships between components that determine whether a technical solution succeeds or fails.

The consciously competent technical seller deliberately seeks out these gaps in the buyer's AI-gathered knowledge. You must listen for opportunities to contribute context, share experiences, offer wisdom, suggest innovation, and identify integration challenges. As AI becomes more sophisticated, these uniquely human contributions become increasingly valuable.

Understanding these gaps isn't enough; we must also change how we approach discovery itself. To truly add value beyond AI, technical sellers must shift from a transactional mindset to one of continuous learning and adaptation. That's where our first principle begins.

The First Principle: Discovery Never Ends

Before we explore specific discovery techniques, we must establish the first principle of Apex 8 discovery: Discovery never ends.

In traditional sales methodologies, discovery is treated as a distinct phase—a box to be checked before moving on to solution presentation. This linear thinking fails in complex technical sales, particularly in the AI era, where information continuously evolves throughout the buying process.

In Apex 8, discovery is not a phase but a mindset that persists throughout the entire sales process. Every conversation, demonstration, and negotiation is an opportunity to uncover new aspects of the buyer's reality. Several later steps in the methodology—particularly Validate and Rehearse—explicitly reopen discovery channels to ensure our understanding remains current and complete.

This continuous discovery approach offers several advantages:

1. **Adaptation to Changing Needs:** Buyer requirements often evolve as they learn more about potential solutions. Continuous discovery allows you to track these changes.
2. **Progressive Revelation:** Buyers rarely reveal all relevant information in initial conversations. Trust builds over time, leading to more candid sharing of challenges and constraints.
3. **Emerging Stakeholders:** New decision-makers and influencers often emerge throughout complex sales processes. Continuous discovery helps identify and address their specific concerns.
4. **Knowledge Refinement:** Your understanding of the buyer's needs becomes more nuanced over time, allowing for more precise solution development.

Here's a truth that's cost me deals: the moment I thought I fully understood a buyer's world was exactly when I stopped asking questions—and invariably, that's when critical insights started slipping right past me. The best sellers I know maintain a healthy paranoia that they're missing something vital, right up until implementation begins (and honestly, even then).

Implementing continuous discovery requires deliberate practices that we'll explore throughout this book. For now, remember that even as you move through subsequent stages of the Apex 8 process, you must maintain an active discovery mindset, constantly refining your understanding of the buyer's reality.

Uncovering the Three Realities

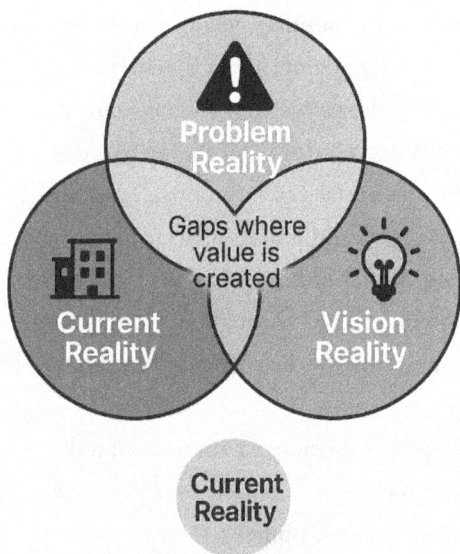

The Three Realities Framework

Problem Reality

Gaps where value is created

Current Reality

Vision Reality

Current Reality

Effective discovery in technical selling means uncovering three distinct but interconnected realities:

1. **Current Reality:** The buyer's present situation.
2. **Problem Reality:** The challenges driving their search for a solution.
3. **Vision Reality:** Their desired future state.

Each reality requires different discovery approaches, and the gaps between these realities create opportunities for your solution. Let's explore each in detail.

Current Reality

Current Reality discovery focuses on understanding where the buyer is today—their existing systems, processes, organizational structure, and business model. In the AI era, much of this information may be publicly available or easily accessible through AI research tools. However, the consciously competent seller goes deeper, exploring:

- **System Landscape:** Beyond just identifying which systems are in place, understand how they interact, their history, and the organizational politics surrounding them.
- **Performance Metrics:** How the organization currently measures success, both formally and informally.
- **Cultural Factors:** Unwritten rules, past experiences with similar initiatives, and cultural attitudes toward change and technology.
- **Resource Realities:** The true availability of budget, time, expertise, and attention—often very different from what appears on organizational charts or financial statements.
- **Decision Dynamics:** How decisions are actually made, regardless of formal processes.

In an engagement with a specialty wood manufacturer in Oregon, I discovered a crucial reality that previous consultants had missed. During an informal lunch, I asked the operations principal a simple question: "If you could have any technology out there, what would make the biggest difference in your company?" His answer: "A forecast optimizer that automatically adjusts as the wood yield varies." This revealed an unexpressed need that only he and I really knew about, leading us to innovate a new type of solution. With yield constantly changing every few days (you don't know what grade wood you really have until you process it), their forecasts needed real-time adjustment. No one had ever mentioned it before because the team assumed it wasn't possible. They didn't know what could be done. Not only did we implement the solution, but the integration framework we built has since allowed them to enhance it further with AI capabilities that now do it much better and faster than ever before.

This exploration of Current Reality must go beyond surface-level facts that AI can easily gather. You're seeking the context, history, and human dynamics that shape how the organization actually operates rather than how it appears from the outside.

Problem Reality

Problem Reality discovery explores the challenges driving the buyer's search for a solution. While AI can identify stated problems from public information, the skilled technical seller uncovers:

- **Root Causes vs. Symptoms:** Distinguishing between the surface issues buyers identify and the deeper root causes.
- **Problem History:** Understanding previous attempts to solve these challenges and why they failed.

- **Problem Perception:** How different stakeholders view the problem and its priority.
- **Problem Impact:** Tracing how the problem affects various parts of the organization, particularly impacts that the buyer may not have recognized.
- **Problem Timeline:** Understanding the urgency drivers and how long the problem has existed.

Problem Reality discovery often reveals that the buyer's stated problem is not their actual problem. The technical seller who only addresses the stated issue may deliver a solution that fails to resolve the underlying challenge, leading to dissatisfaction even if all requirements are technically met.

Vision Reality

Vision Reality discovery explores the buyer's desired future state—what they hope to achieve by implementing a new solution. This is where the greatest opportunity for differentiation exists, as AI struggles to help buyers articulate visions that differ significantly from current patterns. The skilled technical seller explores:

- **Outcome Priorities:** Which business outcomes matter most and how they'll be measured.
- **Success Visualization:** How the buyer imagines success looking and feeling when achieved.
- **Aspiration vs. Expectation:** The gap between what the buyer hopes for and what they actually expect to achieve.
- **Vision Alignment:** How consistent the vision is across different stakeholders.
- **Vision Evolution:** How the buyer's vision has changed over time and what drove those changes.

Vision Reality discovery is particularly powerful because it often reveals that buyers haven't fully articulated their desired future state even to themselves. By helping them develop and clarify this vision, you create value before you've even proposed a solution.

Each of these realities—where the buyer is now, what's holding them back, and where they want to go—offers insight. But the real power lies in understanding the gaps between them. That's where your value lives: guiding them from problem to possibility with clarity no AI can provide.

When exploring these three realities, the key is to put yourself in your buyer's shoes—seeing the situation through their eyes. Consider having your AI agent create a simple chart that breaks down these three realities along with starter questions for each. Ideally, use the same large language model (LLM) or AI assistant that you used when researching the company, allowing you to build on that context. Before the meeting, prompt your AI to generate targeted questions based on your account research, customizing your discovery approach to that specific organization.

During the conversation, use questions that help buyers visualize and articulate their experiences: "What does your current process look like when it's happening?", "How does this problem manifest on a typical Tuesday?", or "When you imagine the ideal outcome, what do you see happening?" This real-time exploration reveals nuances that even the most comprehensive pre-meeting research can't uncover.

Now that we've defined the three layers of buyer reality, let's explore practical techniques that help you uncover them in live conversations—often revealing insights that no AI could surface.

The Art of Technical Discovery

Uncovering these three realities requires moving beyond standard discovery questions to deeper exploration. Here are key techniques for the art of technical discovery in the AI era:

Contextual Inquiry

Rather than relying solely on direct questions, contextual inquiry involves observing buyers in their natural environment and exploring their challenges through conversation. This might include:

- Virtual tours of facilities or systems
- Observation of current processes
- Shadowing users as they perform relevant tasks
- Reviewing actual documentation and artifacts

These approaches reveal realities that buyers themselves may not consciously recognize and thus wouldn't mention in response to direct questions.

Problem Archaeology

Problem archaeology involves tracing issues back to their origins to understand their full context:

- "When did you first notice this challenge?"
- "What was happening in the organization when this issue emerged?"
- "How has the problem evolved over time?"
- "What solutions have you already tried, and what happened?"

This archaeological approach often reveals patterns and underlying factors that explain why previous solutions failed and what a successful approach must address.

Stakeholder Mapping

Complex technical sales involve multiple stakeholders with different priorities, concerns, and levels of influence. Stakeholder mapping helps you understand this landscape:

- Identifying all relevant decision-makers and influencers
- Mapping their formal and informal relationships
- Understanding their individual goals, fears, and motivations
- Assessing their level of support or resistance to change

I'll never forget a deal with a major healthcare network where we spent weeks working with the official buyer—the CTO. It was a perfect technical fit, the budget was approved, and everyone nodded along in meetings. Then suddenly... nothing. Radio silence. It turned out that a mid-level operations director, who barely spoke in meetings, had the CEO's complete trust. Once we identified this hidden influence pattern and brought her deeper into the conversation, addressing her specific concerns about implementation timing, the deal closed within days. That's the power of mapping real influence, not just titles.

This mapping enables you to address the human ecosystem surrounding any technical decision, often the determining factor in whether a solution is adopted successfully.

Future casting

This technique helps buyers explore their desired outcomes by imagining success and then working backward to uncover what it took to get there.

- "Imagine it's one year after implementation. What specific outcomes tell you this has been successful?"
- "Looking back from that future point, what were the key milestones that got you there?"
- "From that future perspective, what obstacles did you have to overcome?"

This approach helps buyers move beyond vague aspirations to specific, measurable outcomes that can guide solution development.

Constraint Surfacing

While most discovery focuses on needs and wants, constraint surfacing deliberately explores limitations that will shape any potential solution:

- Budget realities and approval thresholds
- Technical constraints of existing systems
- Regulatory and compliance requirements
- Timing limitations and deadlines
- Organizational change capacity

When we talk about surfacing constraints, many sellers worry they're limiting opportunity. But here's the truth: identifying constraints early isn't just smart—it builds serious trust. Buyers are exhausted by vendors who promise everything and deliver disappointment. When you proactively say, "Based on what you've shared, these three approaches won't work for you because...", you're signaling that you value their time more than your commission. They instantly know you're a different kind of partner who's playing the long game.

By surfacing these constraints early, you avoid developing solutions that look perfect on paper but prove impossible to implement in the buyer's actual environment.

AI-Augmented Discovery Tools

While we've focused on the limitations of buyer-side AI research, seller-side AI offers powerful tools to enhance human discovery capabilities. The Apex 8 methodology incorporates several AI-augmented discovery approaches:

Signal Detection

AI can analyze buyer communications (with appropriate permissions) to identify signals that might be missed by human observation alone:

- Changes in engagement levels over time (the most obvious telltale)
- Urgency indicators in language patterns (e.g., "quarter end" or "fiscal year" suddenly appearing)
- Inconsistencies in stated requirements (what they say in group versus one-on-one)
- Emotional responses to specific topics (that flash of concern when you mention data migration)
- Alignment or misalignment between stakeholders (the subtle glances when someone makes a claim)

These signals provide insight into underlying concerns, priorities, and dynamics that buyers may not explicitly state.

Pattern Recognition

AI can identify patterns across interactions that reveal deeper truths about the buying organization

- Topics consistently avoided or redirected
- Questions that generate hesitation or uncertainty
- Terms that trigger positive or negative reactions
- Decision patterns that indicate organizational culture

These patterns often reveal unspoken concerns or priorities that shape the buying process.

Influence Mapping

AI tools can help map formal and informal influence networks within buying organizations by analyzing:

- Communication patterns in email and meeting transcripts
- Document workflows and approval sequences
- Reference patterns (who refers to whom and how)
- Meeting dynamics and speaking patterns

This mapping helps identify the true decision structure beyond what appears on organizational charts.

Knowledge Integration

AI can integrate discoveries from multiple touchpoints and team members into a coherent understanding of the buyer's situation:

- Reconciling information from different stakeholders

- Highlighting contradictions for further exploration
- Tracking the evolution of requirements over time
- Identifying gaps in current understanding

This integration ensures that insights aren't lost between conversations and that the full team benefits from each discovery interaction.

When using these AI-augmented discovery tools, the consciously competent seller maintains ethical standards by:

- Being transparent with buyers about data usage
- Focusing on understanding rather than manipulation
- Using insights to create genuine value for the buyer
- Respecting privacy and confidentiality boundaries

The goal is not to use AI to outsmart the buyer but to better understand and serve them through more complete discovery.

Used thoughtfully, these tools empower the consciously competent seller to uncover richer truths—ethically, transparently, and in service of the buyer's success.

Applying Discovery Principles in Practice

Let's conclude with practical applications of these discovery principles in your next client interaction:

Prepare Beyond the Obvious

Before your next discovery conversation:

1. Use AI tools to gather basic information, but recognize that this is just the foundation.

2. Research what AI assistants are likely telling your buyers about your solution.

3. Prepare questions that go beyond what AI can answer—focusing on context, experience, and organization-specific factors.

4. Review your question list and ruthlessly eliminate anything that AI could answer. If ChatGPT can provide information about your solution's specifications, integration capabilities, or typical ROI, don't waste valuable conversation time on those topics. Instead, concentrate exclusively on what only you—a human with experience and pattern recognition—can uncover: the unwritten rules of their organization, the fears that remain unvoiced, and the real priorities that go beyond what's stated in the RFP. That's where the deal actually lies.

Structure for Depth

When conducting discovery conversations:

1. Acknowledge what the buyer likely already knows from AI research.

2. Frame your conversation as a continuation of that foundation—e.g., "I'm sure you've already done some research on this, but I'd really like to know how these options align with your expectations or needs."

3. Move quickly from what to why and how by asking questions like "What did you like about it?", "Why do you think it will fit?", and "How is that different from some of the other products out there, including mine?"

4. Use silence strategically—give buyers time to reflect beyond their prepared responses.

5. Listen for contradictions between what AI would suggest and what the buyer actually says.

Capture Holistically

When documenting discovery findings:

1. Record not just facts but also context and non-verbal signals.
2. Note areas where the buyer's understanding differs from what AI would provide.
3. Document gaps and inconsistencies for further exploration.
4. Track the evolution of requirements and priorities over time.
5. Maintain a living discovery document that continues to evolve throughout the sales process.

Connect to Next Steps

As you prepare to move from Discovery to Prototype:

1. Validate your understanding of the three realities with the buyer.
2. Identify areas requiring further discovery before solution development.
3. Set expectations for continued discovery throughout the process.
4. Share insights that demonstrate the value you've already created through discovery.
5. Preview how these insights will shape your prototype approach.

Remember, discovery in technical selling is not about merely checking boxes on a qualification list. It's about developing a deep understanding of the buyer's realities that goes beyond what AI can provide. This understanding becomes the foundation for everything that follows in the Apex 8 process.

Regarding how much of this you want to apply to any one pursuit, I leave that up to you. I grant you some autonomy here as a salesperson because, ultimately, you must assess how much time and money you want to invest in preparation versus diving right in. However, consider this: it all relates to the level of competition you face, the size of the commission you will earn, and how important it is for both you and your customer to finalize this agreement.

If it's for a fairly commoditized technical item, you probably won't need to do all of this. You may have encountered similar situations so often that you already know the answers before they're given, and I understand that. But I'm presenting the most thorough approach first, and you can simplify or remove certain questions if it pertains to a more straightforward technical product.

However, if you're selling a new AI vertical stack—something that every competitor wants—then all this preparation and execution will be well worth it, especially if it offers a higher commission than some of your other products.

In the next chapter, we'll explore how to transform these discovery insights into the second step of Apex 8: Prototype. We'll examine how to create conceptual models of your solution that address the realities you've uncovered while differentiating your approach from competitors in the AI era.

As we move forward, maintain that discovery mindset. The most valuable insights often emerge when you least expect them, and your willingness to continuously refine your understanding will set you apart from sellers who believe discovery ends when the presentation begins.

CHAPTER 3

PROTOTYPE: CREATING SOLUTION VISIONS WITH INTELLIGENT AUGMENTATION

From Pitching to Prototyping

Elaine surveyed the conference room where six executives from Regional Telecom awaited her presentation. As a senior account executive for NextGen Networks, she had spent the last decade selling advanced telecommunications infrastructure. This opportunity—a comprehensive 5G network upgrade across three states—represented a potential $40 million deal.

In the past, she would have launched into a standard capabilities presentation, detailing NextGen's technical specifications, deployment methodology, and reference clients. But today, she chose a different approach.

"Before I share our thoughts," Elaine began, placing a rolled-up architectural diagram on the table, "I want to confirm what we learned in our discovery conversations." She unveiled a customized network architecture map, highlighting Regional Telecom's existing infrastructure along with proposed 5G integration points marked in blue.

"Based on our discussions, we've created this preliminary network vision. It's not final—it's a starting point for us to refine together."

She pulled out markers and handed them to the CTO. "In fact, I'd like you to mark any areas where you see issues or opportunities we've missed."

The energy in the room shifted immediately. The executives leaned forward, pointing to various sections of the diagram. The CTO began marking areas with questions and comments. Within minutes, the passive audience had transformed into active participants, co-creating a solution rather than merely evaluating a pitch.

This is the power of prototyping in technical sales—the second step in the Apex 8 methodology.

In the AI era, where buyers can access detailed product specifications, comparative analyses, and technical reviews before ever speaking with you, traditional proposals often fall flat. Buyers don't need you to explain what your solution is—AI has already done that. They need you to show how your solution transforms their specific situation in ways they haven't yet imagined.

Prototyping creates this vision through the collaborative development of conceptual models tailored to the buyer's unique context. It transforms the sales conversation from "Here's what we sell" to "Here's what we can build together."

The Timing Revolution: Compressed Sales Cycles

One of the most dramatic shifts in technical selling that is often overlooked is the radical compression of sales cycle timing. What once unfolded over months now often happens within days or even hours. This acceleration demands a fundamental rethinking of when prototyping occurs in the sales process.

In the pre-AI era, the typical sequence was predictable:

1. Initial discovery meeting
2. Follow-up with additional stakeholders
3. Requirements gathering
4. Internal solution development
5. Proposal presentation (often weeks later)
6. Technical review

Today, this linear timeline has collapsed. For many technical sales situations, you must now be prepared to:

- Conduct discovery
- Develop a prototype
- Begin validation

All within the first meeting or certainly by the second interaction.

This sounds impossible by traditional standards—and it would be, without AI assistance. But in 2025, AI-powered prototyping tools have transformed what's possible. While engaged in conversation with the buyer, your AI sales assistant can rapidly develop customized architectural diagrams, value models, or process flows based on the information being shared in real time.

For example, while Elaine discussed network requirements with Regional Telecom, her AI assistant (discreetly running on her tablet) was simultaneously:

- Analyzing their existing network topology
- Generating coverage models based on the conversation
- Creating preliminary ROI projections using industry benchmarks
- Developing alternative deployment scenarios for discussion

By the time the conversation reached the point where a prototype would be valuable, the foundation was already prepared, allowing Elaine to smoothly transition from discovery to collaborative prototyping without the traditional delay.

Now, a word of caution about this speed revolution. Moving fast is great, but not if it means skimming the surface. Speed must never come at the cost of depth—your AI tools should enhance your ability to deliver insights, not provide an excuse to skip the thoughtful work that makes co-creation meaningful. I've seen sellers become so enamored with speed that they start treating their AI tools as replacements for genuine understanding. That's a recipe for disaster. The best sellers let AI handle the mechanical tasks—the renderings, the calculations, the formatting—while they focus their human intelligence on the nuances of what the buyer is really saying, what they're struggling to articulate, and how they're reacting emotionally to different ideas. That's the division of labor that wins deals.

This is not science fiction—it's the current reality of technical selling for leading organizations. If you haven't yet integrated AI-powered prototyping tools specifically designed for your industry into your sales process, this represents both an urgent gap to address and a significant opportunity for competitive advantage.

These industry-specific AI prototyping tools differ dramatically from generic AI. They combine:

- Deep domain knowledge of your specific technical field
- Understanding of standard architectures and approaches
- Industry-specific value models and metrics
- Visualization capabilities tailored to your solution type

For telecommunications, this might be an AI assistant that understands network architecture, spectrum allocation, coverage modeling, and telecommunications economics. For manufacturing, it could be a system that comprehends production line design, throughput optimization, and quality control mechanisms.

The key point is that the compressed timing of modern technical sales makes these tools necessities, not luxuries. Without them, you simply cannot operate at the speed buyers now expect.

Redefining "Prototype" for Technical Sales

When I use the term "prototype" in technical selling, I'm not referring to a working model of a product. I'm describing a conceptual representation of a solution specifically designed for a particular buyer's situation.

There is a significant difference between what I'm discussing and the old-school proposal process. Traditional proposals are polished, locked-down PDFs that get emailed after weeks of work—they scream "take it or leave it." Sales prototypes are living, breathing tools that evolve as you collaborate. Think of proposals as formal declarations from on high, while prototypes are ongoing conversations where ideas are exchanged, assumptions are challenged, and alignment happens naturally. One is a presentation; the other is a partnership.

A sales prototype can take many forms:

- A customized architectural diagram
- A process flow visualization
- A mock-up of user interfaces or experiences
- A value realization model

- A risk mitigation framework
- A narrative describing future operations
- A visual representation of business outcomes

What these formats share is their purpose: to make abstract possibilities concrete and to transform generic capabilities into specific solutions.

The key distinction between prototyping and traditional proposal development is timing and collaboration. In conventional sales processes, sellers gather requirements, disappear to develop formal proposals, and then return to present them. In prototyping, sellers create preliminary solution concepts early in the process and develop them collaboratively with buyers, long before formal proposals are requested.

This approach offers several critical advantages in technical sales:

It In the AI era, where differentiation through product knowledge alone is increasingly difficult, the ability to prototype compelling, buyer-specific solutions becomes a crucial competitive advantage.

The Three Dimensions of Effective Prototypes

However, not every prototype is worth the pixels it's displayed on. If you want your prototype to genuinely influence a technical sale, it must address the buyer's concerns from every angle. You can't just focus on technical aspects and call it a day. You need to encompass the full spectrum—technical feasibility, business impact, and the human side of implementation. Miss any one of these, and watch how quickly your impressive prototype gets filed in the "interesting but not actionable" folder.

Effective prototypes in technical sales address three essential dimensions of the buyer's reality. Neglecting any one dimension creates a fatal flaw that can derail even the most technically sound solution.

The Technical Dimension

The technical dimension addresses feasibility, functionality, and performance. It answers questions like:

- Will this solution work in our specific environment?
- How will it integrate with our existing systems?
- Will it deliver the performance we need?
- How will it scale as our needs evolve?

In our telecommunications example, Elaine's network architecture map addressed this dimension by showing specifically how NextGen's 5G technology would integrate with Regional Telecom's existing infrastructure.

Technical prototypes might include:

- Architecture diagrams showing integration points
- Capacity models demonstrating performance under various conditions
- Technology compatibility assessments
- Technical risk analyses with mitigation approaches

The mistake many technical sellers make is stopping at this dimension. While technical feasibility is necessary, it is far from sufficient for complex sales.

The Business Dimension

The business dimension addresses value, outcomes, and economics. It answers questions like:

- What business outcomes will this solution enable?
- How will it affect our key performance indicators?
- What is the expected return on investment?
- How will it impact our competitive position?

Business dimension prototypes might include:

- ROI models customized to the buyer's specific situation
- Business process maps showing efficiency improvements
- Market opportunity analyses enabled by the solution
- Comparative economic scenarios with and without the solution

I love this example from a SaaS sale I was coaching. The rep built a value model showing how the platform would reduce customer churn by 15%—which for this particular company meant about $8 million in retained revenue. The beautiful part? The CFO grabbed that exact model, barely changed a thing, and used it as the centerpiece of his board presentation. Why? Because the model was transparent—every assumption was visible, and every calculation traceable. The board grilled him on the numbers, and he could defend each one because he understood the model inside and out. That deal got greenlit with minimal friction because what started as our prototype became their business case. The CFO didn't have to build anything from scratch—we did the heavy lifting, he did the validation, and everyone won.

For Regional Telecom, Elaine might include a market coverage map showing new service areas enabled by the 5G deployment, along with projected revenue potential for each zone.

The Human Dimension

The human dimension addresses adoption, change management, and user experience. It answers questions like:

- How will this solution affect the people who use it?
- What changes in behavior or process will be required?
- How will we manage the transition from the current to the future state?
- What resistance might emerge, and how will we address it?

Human dimension prototypes might include:

- User experience mock-ups
- Training and change management plans
- Organizational impact assessments
- Adoption timeline scenarios

For the telecommunications deployment, this might include visual representations of how field technicians would install and maintain the new equipment or how customer service representatives would leverage the enhanced network capabilities to better serve subscribers.

The most powerful prototypes address all three dimensions, creating a comprehensive vision of how the solution will work technically, deliver business value, and succeed in human implementation. This three-dimensional approach directly counters the limitations of AI-generated analyses, which typically excel at technical comparisons but struggle with business and human factors specific to a particular organization.

Prototype Formats for Complex Solutions

Different selling situations call for different prototype formats. The consciously competent seller selects formats that best address the buyer's primary concerns and decision-making style.

Here's a practical tip: Take 15 minutes right now and create a simple table that maps out these formats by audience, use case, and core strengths. I guarantee it'll become one of your most-referenced tools when planning which prototype approach to use in your next complex sale. We're all drowning in information—a quick-reference guide like this will save you tons of mental bandwidth when you're under pressure.

Conceptual Architecture Diagrams

Best for: Technical decision-makers focused on infrastructure, integration, and system relationships

Architecture diagrams visualize how your solution fits within the buyer's technical environment. They show integration points, data flows, system relationships, and technical components.

In telecommunications, this might map how new 5G equipment connects to existing network infrastructure, identifying backhaul requirements, spectrum utilization, and coverage patterns.

The power of architecture diagrams comes from their ability to make complex technical relationships visible and discussable.

They transform abstract integration challenges into concrete visual representations that both technical and non-technical stakeholders can understand.

Process Flow Visualizations

Best for: Operations-focused buyers concerned with workflow, efficiency, and procedural changes

Process flow visualizations show how business processes will change with your solution implemented. They map the current state, transition state, and future state of key workflows.

For a telecommunications provider, this might visualize the subscriber provisioning process before and after implementing new network automation tools, showing a reduction in steps and time required.

Process flows are particularly effective for demonstrating operational improvements and highlighting inefficiencies in current approaches that buyers may have accepted as inevitable.

Value Realization Models

Best for: Financial decision-makers focused on ROI, cost justification, and business outcomes

Value realization models quantify the expected business impact of your solution in the buyer's specific context. They go beyond generic ROI calculators to incorporate the buyer's actual metrics, timeframes, and value drivers.

For Regional Telecom, this might model subscriber growth, reduced churn, increased average revenue per user, and new service opportunities enabled by 5G deployment—all based on their specific market demographics and competitive position.

The most effective value models are transparent, allowing buyers to adjust assumptions and see how changes affect outcomes. This transparency builds credibility and allows buyers to take ownership of the economic case.

Experience Mock-ups

Best for: User-focused buyers concerned with adoption, usability, and stakeholder acceptance

Experience mock-ups visualize how users will interact with your solution. They can range from simple interface sketches to interactive simulations, depending on the complexity of the solution.

Here's a perfect example from a telecom deal I was involved with. The technical team had been battling for months with a super skeptical VP of Customer Experience who was blocking an entire system upgrade. Her concern? Training time for her already overloaded call center team. We were getting nowhere with specs and capabilities decks. Then one of our sellers did something brilliant—he sketched out a simple mock-up showing exactly what the agent's screen would look like and how they could troubleshoot common customer issues with literally one click. The VP stared at it for a minute and said, "That's ALL they need to do? Just this one screen?" You could literally see the lightbulb moment. All that resistance melted away because she finally SAW the path to adoption that had been completely missing from our feature-heavy technical presentations. One simple sketch accomplished what months of meetings couldn't.

In telecommunications, this might show how network operations center personnel would use new monitoring dashboards or how field technicians would access diagnostic tools on mobile devices.

These mock-ups make abstract capabilities tangible and help buyers envision the human experience of using your solution, addressing adoption concerns that often derail technically sound implementations.

Narrative Scenarios

Best for: Strategic buyers focused on future vision, competitive advantage, and transformational impact

Don't think of narrative scenarios as reports—think of them as the storyboards for the movie of your buyer's future. When done right, they're cinematic—rich with detail, emotionally relatable, and still firmly rooted in business reality. A great narrative scenario doesn't just inform the logical brain; it immerses the whole person in a future they can almost touch. It's the difference between reading about a vacation destination and smelling the salt air while feeling sand between your toes.

Narrative scenarios tell the story of how the buyer's organization will operate with your solution in place. They paint a vivid picture of the future state, describing specific situations, challenges, and outcomes.

For a telecommunications provider, this might describe how they'll respond to a competitor's move, launch a new service, or handle a natural disaster—all enabled by the capabilities of the new network infrastructure.

Effective narratives connect technical capabilities to strategic out-comes in ways that engage both analytical and emotional decision-making, helping buyers "live in the future" before committing to creating it.

Risk Reduction Frameworks

Best for: Security, compliance, or risk-averse buyers

Risk reduction frameworks identify potential implementation challenges and their mitigations. They demonstrate foresight and expertise by addressing concerns before they arise.

For a telecommunications network, this might involve mapping security vulnerabilities, potential service disruptions during tran-sition, regulatory compliance challenges, and their respective mit-igation strategies.

These frameworks transform vague anxieties into specific risks with defined solutions, reducing the perceived danger of change and showcasing your expertise in managing complex implementa-tions.

Another practical suggestion: when creating your own version of this material for your team, add visual markers to enhance skim-mability. Use icons or bold highlights to quickly identify which formats work best for various stakeholders. Your brain processes visual signals faster than text, and when prepping for a big meeting with multiple stakeholders, you'll appreciate this time-saver. Being able to scan and think, "CTO coming? Architecture diagram. CFO attending too? Bring the value model," can make all the dif-ference.

Trust-Building Progression

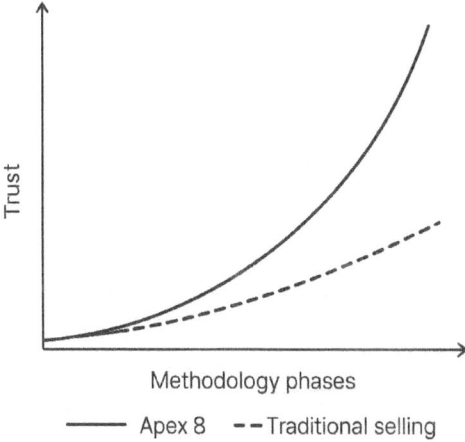

Let me share something I took years to fully appreciate. While functionality and formats are important, they miss the psychological magic happening beneath the surface. Prototyping isn't just about demonstrating what your solution can do—it's a trust accelerator that transforms the buyer-seller dynamic. In a world where trust typically takes months to build, effective prototyping can dramatically collapse that timeline.

While we've explored the functional aspects of prototyping, it's essential to recognize its most powerful effect: the rapid development of authentic trust between seller and buyer. At its core, technical selling isn't just about solutions—it's about relationships built on trust.

Prototyping accelerates trust development in ways traditional selling approaches cannot match:

Vulnerability Creates Connection

Presenting a work-in-progress prototype instead of a polished final product demonstrates vulnerability. You're saying, "This isn't perfect yet—I need your input to make it better." This vulnerability signals confidence in your expertise while showing respect for the buyer's knowledge. It transforms the dynamic from seller versus buyer into collaborators working toward a shared goal.

For Elaine at Regional Telecom, unveiling an incomplete network architecture diagram and explicitly asking for feedback showcased both competence (she had done substantial work) and humility (she valued their expertise). This combination builds trust far faster than presenting a seemingly perfect solution that allows no room for buyer contribution.

Competence Made Visible

Trust in technical selling has two components: character trust (believing you have the buyer's best interests at heart) and competence trust (believing you have the capability to deliver). Prototyping makes your competence visible and tangible in real-time.

When buyers see you rapidly adapt a prototype based on their feedback, incorporate complex requirements into a coherent solution, or collaboratively navigate technical challenges, they experience your competence directly rather than just hearing claims about it. This demonstrated expertise builds confidence in your ability to execute the full solution.

Authentic Over Artificial

Buyers are inundated with glossy, AI generated presentations that all look the same and feel utterly soulless. What they crave is that refreshing moment when someone assesses their specific situation and confidently states, "Actually, that approach won't work here, and here's why." That kind of straightforward honesty feels rare today; when buyers encounter it, they do a double-take. It's human, it's helpful, and most importantly, it feels real in a sea of artificial perfection.

In an era where AI can generate seemingly flawless presentations and proposals, buyers increasingly value authentic human interaction. Prototyping fosters authenticity through real-time problem-solving, creative adaptation, and genuine collaboration.

Even if you use AI tools to generate aspects of your prototype (as discussed in our timing revolution section), the collaborative development process remains deeply human. This human element establishes the foundation of trust necessary for complex technical sales.

Team Leadership Visibility

When prototyping involves multiple team members, it provides a powerful opportunity to showcase your leadership capabilities. How you coordinate resources, leverage team expertise, and maintain direction throughout the collaborative process demonstrates to buyers not just what your solution can do, but how effectively you can manage its implementation.

Even when you bring in specialists to contribute to specific aspects of the prototype, maintaining clear ownership of the process reinforces that you're the trusted guide through this journey. Buyers will associate the value created during prototyping with your leadership, strengthening their confidence in working with you long-term.

The Trust Maintenance Imperative

It's crucial to understand that the trust built through effective prototyping requires continuous reinforcement. Each interaction either strengthens or weakens the foundation you've established, creating a trust maintenance imperative that extends through all subsequent steps of the Apex 8 methodology.

The consciously competent seller recognizes that trust, once granted, must be continually earned through consistent demonstrations of competence, character, and authentic concern for the buyer's success. Every validation activity, rehearsal decision, and presentation element should reinforce the trust established during prototyping.

In many ways, prototyping is the inflection point in technical selling—the moment when the relationship transforms from transactional to trusted advisory. This accelerated trust development becomes the foundation upon which all subsequent steps build, making it perhaps the most crucial psychological aspect of the entire Apex 8 methodology.

Techniques for Collaborative Prototyping

Effective co-creation requires specific techniques that actively engage buyers in the development process:

The Incomplete Solution

Rather than presenting a finished prototype, deliberately leave aspects incomplete or marked with questions. This creates natural entry points for buyer contribution.

When Elaine unveiled her network architecture diagram, she marked several areas with question marks and notes like "Alternative approach?" or "Confirm capacity requirements here." These deliberate gaps invited the Regional Telecom team to fill them.

The Alternative Options Approach

Instead of using the standard "here are my options" approach, try what I call the Alternative Options play. Present two or three fully credible ways to tackle a key challenge—each with clear pros and cons—and genuinely ask the buyer which path makes more sense for them. Watch what happens. The whole dynamic shifts. You're no longer the vendor pushing your one perfect answer; you're a strategic advisor helping them navigate tricky decisions. That's positioning gold.

Elaine might present three different approaches to handling backhaul requirements, each with different trade-offs in cost, performance, and implementation complexity, then facilitate a discussion on which best meets Regional Telecom's needs.

The Real-Time Modification Session

Conduct working sessions where you modify prototypes in real-time based on buyer feedback. This demonstrates responsiveness and creates a powerful sense of progress.

Using digital tools, Elaine could adjust network coverage models during the meeting as executives shared information about priority service areas, immediately showing how these changes would affect overall design and performance.

The Stakeholder Insight Collection

Circulate preliminary prototypes to various stakeholders before key meetings, collecting their insights and incorporating them into revised versions. This broadens buy-in and ensures diverse perspectives are considered.

Before meeting with Regional Telecom's executive team, Elaine might share elements of her network design with their field operations manager, network security director, and customer experience team, incorporating their feedback into the version she presents to executives.

Navigating Co-Creation Challenges

Co-creation isn't without challenges. The consciously competent seller anticipates and addresses these potential pitfalls:

Managing Divergent Input

What happens when your buyer wants to steer your prototype in a technically unsound direction? This occurs frequently. Someone might say, "Can't you just add this capability?" and you know it would create a house of cards. Traditional sales training would advise you to "overcome the objection" or use manipulative techniques to redirect them. Don't do that. Instead, get curious. Ask them, "Help me understand what need you're trying to address with that approach?" Nine times out of ten, what they're asking for isn't what they actually need—it's just the only solution they can imagine. Once you understand the underlying need, you can often address it in a way that works technically. This maintains solution integrity while still honoring their input.

Encouraging Quiet Contributors

Have you ever noticed how some people never speak up in group settings, even when they have critical information? These individuals are often your most thoughtful buyers—the ones who process internally before speaking. If you're only listening to the loudest voices in the room, you're missing valuable insights. Create multiple channels for input—one-on-ones, anonymous feedback forms, shared documents where people can comment privately. I've seen deals completely transform when the quiet engineering director finally shares an implementation concern they've been mulling over for weeks but couldn't voice in meetings. Sometimes, your most critical insights come from those who haven't said a word.

Balancing Stakeholder Perspectives

Let's discuss the minefield of organizational politics during co-creation. You'll encounter situations where the Head of Operations advocates one approach while the CTO pushes for something entirely different—and they're both staring you down, waiting to see whose side you'll take. This is where amateurs crash and burn. Never pick sides. Your role is to be the neutral party who focuses on evaluation criteria rather than personalities. Say something like, "Both approaches have merit—let's put them side by side and assess how each performs against your requirements." Document everyone's viewpoints so they feel heard. You're not there to settle their power struggles; you're there to find the solution that works best for the organization as a whole.

Managing Expanding Needs

One of the biggest risks in prototyping is what I call "scope creep on steroids." The more you collaborate with buyers and show them what's possible, the more their appetite grows. Suddenly, they're asking if you can also solve world hunger while you're at it. This isn't greed—it's natural excitement as they see possibilities. But if you don't manage it, it can lead to disaster. I always keep a visible "parking lot" for future-phase ideas. Acknowledge the value of every suggestion, but be diligent about sorting them into "core solution" versus "future opportunity." Make this categorization transparent and collaborative. You're not saying no; you're creating a roadmap that starts with a focused, achievable foundation.

By managing these challenges skillfully, you transform the prototype from a sales tool into a collaborative platform that builds relationships and commitment throughout the buying process.

AI-Augmented Prototyping Tools

Now, let's talk about AI—not in that "AI will take your job" way that makes everybody nervous. As we've seen throughout the Apex 8 process, AI isn't replacing the human element; it's supercharging it. Think of AI as your prototype production assistant on steroids. You're still the director, the creative mind, the relationship builder—AI just helps you execute faster and explore more possibilities than you could on your own. Let me show you how to leverage AI to prototype smarter without losing the authentic human touch that buyers are absolutely craving.

Rapid Visualization Tools

AI-powered visualization tools can transform text descriptions into visual representations—architecture diagrams, process flows, interface mock-ups—in minutes rather than days. This enables real-time creation and modification of visual prototypes during buyer interactions.

For telecommunications infrastructure, AI visualization tools might generate coverage maps, network topology diagrams, or capacity models based on parameters you define with the buyer.

Scenario Modeling Engines

AI can quickly generate multiple scenarios based on different assumptions or approaches, allowing you to explore alternatives with buyers efficiently. These engines can model technical performance, business outcomes, or implementation timelines under various conditions.

For Regional Telecom, an AI scenario engine might model network performance and economic return across different deployment strategies—urban-first versus rural-first, or aggressive versus conservative timelines.

Customization Accelerators

AI can rapidly customize generic solution frameworks to specific buyer contexts by analyzing discovery data. These tools transform standard approaches into buyer-specific starting points for co-creation.

By analyzing Regional Telecom's network data, market position, and strategic priorities, AI could customize baseline network architectures to address their specific challenges before Elaine's first prototyping session.

Insight Aggregators

AI can analyze feedback from multiple stakeholders, identifying patterns, contradictions, and priorities that might not be apparent from individual reviews. This helps synthesize diverse input into coherent direction for prototype refinement.

After collecting input from various Regional Telecom stakeholders, an AI aggregator might identify that network security emerges as a top concern across departments, even when not explicitly stated, suggesting this should be emphasized in the next prototype iteration.

Using AI Ethically in Prototyping

When I discuss ethics in AI-assisted selling, I'm not just checking a corporate compliance box. Let's be clear about how to be both effective AND ethical:

Tell Verify Use Protect

From Prototype to Validation

Effective prototypes naturally lead to the next step in the Apex 8 methodology: Validation. As buyers engage with your prototypes, they'll raise questions, concerns, and requirements that must be validated before proceeding to formal proposals.

Design your prototypes with validation in mind by:

Including Highlighting Defining Creating For Regional Telecom, Elaine might conclude the prototyping phase by proposing specific validation activities: a technical review with their network engineering team, site surveys of key tower locations, and perhaps a limited proof-of-concept deployment in a high-priority market.

A prototype is ready for validation when:

- Key stakeholders have contributed to its development.
- Major questions and concerns have been identified.
- The solution concept addresses all three dimensions (technical, business, human).
- Natural next steps for testing assumptions are apparent.
- Buyers are actively engaged in refining the approach rather than just evaluating it.

Look, the brutal truth about selling today is that you're up against unprecedented buyer skepticism, exhaustive research, and attention spans thinner than ever. This environment demands not just insight but co-created clarity that cuts through the noise. A great prototype doesn't merely describe a hypothetical future state; it pulls your buyer into that future, allowing them to explore, evaluate, and feel ownership before they've spent a dime. That visceral experience of co-creation is something your competitors can't match with their generic slideshows and canned demos. This is your edge—use it.

Before we leave this chapter, let me offer some practical advice that you can apply immediately to your next sales opportunity. What's the fastest way to get started with prototyping if you've never done it before?

Start Simple But Specific

For your very next opportunity:

1. Pick just one aspect of your solution that would benefit from a visual representation—don't try to prototype everything at once.
2. Create a simple, buyer-specific visualization focused on that single aspect—keep it rough enough that it's obviously not a finished proposal.
3. Deliberately leave portions incomplete or marked with questions like "capacity needs verification" or "integration approach TBD."
4. Share it early—ideally in your second conversation, not six meetings in.
5. Bring the right tools so you can modify it on the fly based on feedback—nothing builds credibility faster than making changes in real-time.

Create Multi-Dimensional Prototypes

As you become more comfortable with the basics:

1. Ensure you're addressing all three dimensions we discussed—technical feasibility, business outcomes, and human adoption.
2. Match your prototype format to the decision-maker—an architecture diagram for the CTO, a value model for the CFO, an experience mock-up for end users.

3. Connect technical capabilities directly to business outcomes—show exactly how feature X delivers outcome Y.
4. Don't skimp on the human dimension—demonstrate what implementation and adoption actually look like.
5. Create different views of the same solution for different stakeholders—the operations team needs to see something different than the executive team.

I can't stress this enough: those three dimensions are not optional extras. Technical feasibility, business outcomes, and human adoption must be front and center in your mind with every prototype you create. I see too many sellers who excel in the technical dimension but falter when it comes to business impact or human factors. Your prototype is only as strong as its weakest dimension. Don't be the genius engineer with a technically perfect solution that nobody wants to use and that doesn't make business sense. Hit all three dimensions every single time.

Master Co-Creation

This is where the real magic happens:

1. Ask far more questions than you make statements when presenting prototypes—"How does this approach align with your security requirements?" is infinitely better than "Here's our security architecture."
2. Use phrases like "Based on what we've discussed so far..." to consistently reinforce that this is a work in progress, not a finished solution.

3. Ensure everyone can participate—bring extra copies, tablets, or whatever it takes so multiple stakeholders can mark up simultaneously.

4. Document everything, even seemingly minor comments—that throwaway remark about system latency might turn out to be critical later.

5. When you return with the next version, clearly show where you've incorporated their feedback—"You mentioned concerns about the API throttling limits, so we've redesigned this component to..."

Leverage AI Appropriately

Here's how to use AI in prototyping without losing your human edge:

1. Identify the repetitive grunt work that's slowing you down—generating basic diagrams, crunching numbers for ROI models, or formatting mockups—and let AI handle just those parts.

2. Use AI to quickly generate multiple approaches to the same problem, giving you options to discuss with the buyer.

3. Feed AI the specific context from your discovery to customize standard architectures to the buyer's unique situation.

4. Always review any AI-generated content before sharing it—one hallucinated feature could undermine your credibility.

5. Focus on the strategic parts of prototyping that AI can't touch—reading the room, sensing unstated concerns, building relationships, and navigating organizational politics.

Remember, effective prototyping isn't about impressive visuals or perfect presentations. It's about creating a collaborative platform that transforms the buyer from evaluator to co-creator, building commitment through involvement long before formal proposals are requested

In the next chapter, we'll explore the third step in the Apex 8 methodology: Validate. We'll examine how to confirm that your co-created solution truly addresses the buyer's needs before proceeding to formal presentation.

As you move forward, maintain both the discovery mindset from Chapter 2 and the collaborative approach to solution development we've explored here. The combination of deep understanding and co-created solutions creates a foundation for technical sales success that AI alone cannot replicate.

We've entered a world where anyone can use AI to produce impressive presentations and slick pitches, but nobody has figured out how to make AI inspire genuine confidence or create that spine-tingling moment when a buyer sees their future clearly for the first time. That's still on you. Your ability to co-create—to roll up your sleeves with buyers and build something meaningful together—is what separates order-takers from deal-makers. This human art of prototyping not only sets you apart in a sea of sameness but also builds a foundation of trust that will carry you and your buyer forward together long after the contract is signed.

CHAPTER 4

VALIDATE: CONFIRMING SOLUTION FIT IN A MACHINE-LEARNING WORLD

When Confidence Isn't Enough

Dr. Maya Chen leaned forward in her chair, studying the reactions of the hospital procurement committee. As a technical sales specialist for Precision Diagnostics, she had spent the past hour walking through her prototype for a new advanced imaging system that promised to revolutionize early detection of neurological disorders. The prototype was impressive—customized to the hospital's specific patient population, workflow, and existing infrastructure. The committee members seemed engaged, asking thoughtful questions and even contributing suggestions.

But Maya recognized the moment they had reached: that critical juncture where interest meets doubt. The Chief Medical Officer's next question confirmed it.

"This all looks promising in theory, Dr. Chen," he said carefully. "But we're talking about a seven-figure investment that will affect diagnostic capabilities for thousands of patients annually. How can we be confident this will perform as described in our specific environment, with our particular case mix and staff expertise?"

This is the validation challenge in technical selling—the moment when a compelling prototype collides with real-world skepticism. It's the third step in the Apex 8 methodology and arguably the most crucial for complex technical sales.

In the AI era, buyers have grown increasingly skeptical of claims and projections. They've been burned too many times by solutions that looked perfect in PowerPoint but failed in practice. AI-generated analyses and competitor comparisons have created an abundance of information but a scarcity of trust. Every technical claim is immediately suspect; every promised outcome demands verification.

Maya didn't respond with a typical case study or generic reference. Instead, she opened her tablet.

"I anticipated this question," she replied. "In preparation for today, we worked with Dr. Ramirez at Metro Medical, which has a similar patient demographic and clinical workflow to yours. They've been using our system for six months on cases comparable to yours."

She displayed a dashboard showing anonymized patient data, diagnostic accuracy rates, and workflow metrics from the reference site alongside projected metrics for this hospital.

"But more importantly, we've arranged for your radiology team to remotely operate our system on test cases this Friday. Your staff, your protocols, real diagnostic scenarios—all without disrupting current operations. We'll measure actual performance in your environment before you make any commitment."

The committee's energy shifted perceptibly. They weren't just evaluating a vendor claim anymore; they were preparing to verify real-world performance relevant to their specific needs.

This is validation in the Apex 8 methodology—not just providing evidence but creating experiences that transform skepticism into confidence.

Beyond Traditional Validation

Traditional validation is dead on arrival in today's environment. Those glossy case studies, carefully coached reference calls, and perfectly scripted demos just don't cut it anymore. Today's buyers see through that immediately.

Traditional validation relies on curated stories and ideal demos, but today's buyers see through those. The Apex 8 approach shifts from storytelling to shared experience—verifying fit in the buyer's real environment with their real people.

Think about it: case studies have become suspect because everyone knows they are cherry-picked success stories. With AI content generation, buyers can't even tell if your case studies involve real customers anymore. Reference calls have lost their impact because buyers know they are only speaking to your biggest fans, who have been prepped thoroughly. And those controlled product demos? Please. Buyers can differentiate between your pristine lab environment and their messy reality.

What works now is creating verification experiences that directly address your buyers' specific concerns about implementation in their unique environments. I'm talking about:

- Testing your solution against their actual use cases, not generic capabilities
- Replicating their real-world conditions, not ideal scenarios
- Getting their own team members hands-on with your solution
- Tackling technical, business, and human fit all at once
- Building shared experiences that create genuine confidence on both sides

For Maya at Precision Diagnostics, this meant creating a custom validation experience that would test the hospital's specific concerns about the imaging system's performance with their actual patient population and staff. This approach bridges the massive trust gap that exists in complex technical sales today, especially when selling something with AI or advanced technology components.

The Three Dimensions of Validation

Effective validation in technical sales must address three essential dimensions of solution fit. Neglecting any one of these creates a failure point that often emerges only after the sale, damaging both implementation success and customer relationships.

Technical Validation

Technical validation confirms that the solution will function as promised in the buyer's specific environment. It answers questions like:

- Will it integrate with our existing systems?
- Can it handle our volume and performance requirements?

- Will it function with our specific configurations and work-flows?
- How will it respond to our edge cases and exceptions?
- What technical resources will implementation require?

For Maya's imaging system, technical validation might include compatibility testing with the hospital's PACS (Picture Archiving and Communication System), performance evaluations with their specific image volume and types, and integration testing with their electronic health record system.

Technical validation methods might include:

- Limited pilot implementations
- Integration simulations
- Load testing with buyer-specific data volumes
- Technology architecture reviews
- API and interface testing

The key to technical validation is specificity—testing the solution against the buyer's actual technical environment rather than generic benchmarks.

Business Validation

Business validation verifies that the solution will deliver the expected value and return on investment in the buyer's specific context. It addresses questions like:

- Will we achieve the projected efficiency improvements?
- How will this solution affect our key performance indicators?
- What is the realistic timeline for value realization?
- How do the total costs compare to the expected benefits?

- What business risks might affect successful implementation?

For the hospital considering Maya's imaging system, business validation might include financial modeling based on their actual patient mix, reimbursement rates, and diagnostic volumes. It might also include an analysis of potential revenue from new diagnostic capabilities or reduced liability from improved accuracy.

Business validation methods might include:

- Value realization modeling with buyer-specific inputs
- Sensitivity analysis for different implementation scenarios
- Total cost of ownership calculations
- Business process simulations
- Risk-adjusted ROI projections

Effective business validation goes beyond generic ROI calculators to incorporate the buyer's actual metrics, constraints, and objectives.

Human Validation

Human validation ensures that the solution will be successfully adopted and used by the people in the buyer's organization. It explores questions like:

- Will our staff embrace or resist this solution?
- How will it affect workflows and job satisfaction?
- What training and change management will be required?
- How will it impact the experience of our customers/patients/users?
- What cultural factors might affect implementation success?

For the hospital, human validation of the imaging system might include usability testing with their radiologists and technicians, workflow impact analysis, and assessment of training requirements based on their staff's current skill levels.

Human validation methods might include:

- User experience testing with the buyer's actual staff
- Workflow shadowing and impact analysis
- Organizational readiness assessments
- Change management planning
- Cultural compatibility evaluation

Human validation recognizes that even technically sound solutions with clear business value can fail if they don't align with the human reality of the organization implementing them.

Validation Methods for Complex Solutions

Each validation method is ultimately about reducing the buyer's specific doubts—whether technical, financial, or cultural—so that confidence can replace caution. The right method depends entirely on what keeps your buyer up at night.

Different technical sales situations require different validation approaches. The consciously competent seller selects methods that address the buyer's specific concerns and decision-making process.

Tailored Proof-of-Concept

When in-person testing is possible, a limited real-world rollout offers irreplaceable insight into how the solution performs under actual operating conditions.

Best for: High-risk, high-investment solutions with significant technical uncertainty.

A tailored proof-of-concept (POC) implements a limited version of the solution in the buyer's actual environment to validate specific capabilities or outcomes. Unlike generic demonstrations, a well-designed POC:

- Focuses on the buyer's highest-priority use cases
- Uses the buyer's actual data and environment
- Involves the buyer's team in execution
- Tests against predefined success criteria

For Maya's imaging system, a tailored POC might involve implementing the system for a specific diagnostic procedure on a limited set of cases, allowing the hospital's radiologists to compare results with their current methods.

POCs require careful scoping to be manageable while still providing meaningful validation. They should be designed to verify the aspects of the solution that create the most uncertainty or concern for the buyer.

Reference Alignment Strategy

Best for: Solutions with established track records but questions about applicability to the buyer's specific situation.

Reference alignment goes beyond traditional reference calls by identifying existing customers with environments, challenges, and objectives similar to the prospective buyer. This approach:

- Matches references based on relevant attributes, not just willingness to advocate

- Facilitates structured discussions focused on specific validation questions
- Creates peer-to-peer dialogue rather than vendor-mediated testimonials
- Allows for direct exploration of implementation challenges and lessons learned

Maya's reference alignment strategy involved finding a hospital with similar patient demographics, technical infrastructure, and clinical workflows that had successfully implemented the imaging system. This aligned reference could speak directly to the implementation challenges this new prospect would likely face.

The key to effective reference alignment is specificity—finding existing customers who share meaningful characteristics with the prospect rather than simply showcasing your happiest clients.

Simulation and Modeling

When real-world testing isn't feasible, simulation offers a virtual alternative that mirrors your buyer's environment to predict likely outcomes at scale.

Best for: Solutions where physical implementation for validation is impractical due to scale, cost, or disruption.

Simulation and modeling techniques create virtual environments that replicate the buyer's conditions to predict solution performance. Modern simulation approaches:

- Incorporate actual data from the buyer's environment
- Model specific workflows and processes
- Allow for variable inputs to test different scenarios

- Visualize outcomes in ways that non-technical stakeholders can understand

For the imaging system, Maya might use simulation to model patient throughput under different configurations, predict diagnostic accuracy improvements based on the hospital's actual case mix, or visualize workflow changes without disrupting current operations.

The effectiveness of simulation depends on the quality of the data inputs and model assumptions. The consciously competent seller ensures transparency about these factors rather than presenting simulations as guaranteed predictions.

Expert Validation Panels

Best for: Highly specialized or technical solutions requiring domain expertise to evaluate.

Expert validation panels bring together respected authorities to assess a solution's suitability for a specific buyer. These panels:

- Include independent experts not employed by either buyer or seller
- Evaluate evidence using structured assessment frameworks
- Document findings and recommendations for stakeholders.
- Address both capabilities and implementation considerations.

In one instance, a panel of independent academic researchers issued a findings report validating a biotech system's methodology, enabling the purchasing hospital to fast-track regulatory approval and secure board support. This endorsement reduced the approval timeline by six months and helped neutralize a competitor attempting to cast doubt on the approach.

For medical equipment like Maya's imaging system, an expert validation panel might comprise radiologists from academic medical centers, medical physicists, and healthcare technology assessment specialists who can evaluate the system's clinical and technical merits specifically for the hospital's needs.

The credibility of expert validation hinges on the qualifications of the experts and their independence. A consciously competent seller ensures that the panel includes respected figures without conflicts of interest.

Stakeholder Workshops

Best for: Solutions impacting multiple departments or functions.

Stakeholder workshops gather representatives from various affected areas to assess solution fit through structured activities. These workshops:

- Include participants from all impacted departments or functions.
- Use structured exercises to explore implementation scenarios.
- Document concerns, requirements, and insights.
- Build consensus around solution fit.

In one case, a workshop revealed that IT staff were concerned about bandwidth limitations that had previously gone unmentioned—prompting the sales team to incorporate a network upgrade into the revised proposal. This potentially deal-killing issue would not have surfaced in standard technical reviews, but the workshop format allowed the network administrator to voice concerns he had been keeping to himself.

For the hospital considering the imaging system, a stakeholder workshop might include radiologists, technicians, IT staff, administrators, and finance representatives working through implementation scenarios and identifying potential issues.

The value of stakeholder workshops extends beyond validation to building buy-in and preparing for implementation. They foster shared understanding and ownership that supports successful adoption if the sale proceeds.

Consider creating a quick-reference table summarizing the methods above—listing each method, when to use it, and which type of stakeholder it best serves. You'll frequently refer back to this when planning validation strategies for different opportunities.

The AI Knowledge Trust Gap

As more solutions incorporate AI components, a new type of skepticism arises—one that traditional validation methods often fail to address.

A significant validation challenge has emerged in the AI era, particularly for solutions that utilize artificial intelligence capabilities. I term this the "AI Knowledge Trust Gap"—the growing skepticism buyers have toward algorithm-based systems whose decision-making processes are not fully transparent.

This trust gap is especially pronounced in healthcare, where diagnostic systems using AI undergo intense scrutiny. Buyers pose questions such as:

- How was the AI trained, and on what data?
- Can we verify the algorithm's decision process?
- Will it perform consistently across our patient demographics?
- How do we validate outputs we can't independently verify?
- What happens when the AI makes a mistake?

Maya confronted this challenge directly with Precision Diagnostics' imaging system, which incorporated machine learning for certain diagnostic procedures. Her validation approach needed to address not only whether the system worked but also how it worked and why it could be trusted.

To bridge this gap, sellers must provide new forms of validation—ones that render AI performance transparent, trustworthy, and verifiable within the buyer's context.

Strategies for Validating AI-Powered Solutions

When selling solutions with AI components, consider these specialized validation approaches:

Demystify the Algorithm

- Transparent Data sources used for training.
- Validation methods employed during development.
- Known limitations and edge cases.
- Ongoing learning and update processes.

This documentation should be accessible to both technical and non-technical stakeholders, emphasizing understanding over technical details.

Explainable Think of explainability as showing your work on a test—buyers want not just the answer, but also to see how the AI derived it, and whether the logic holds up under scrutiny. If your AI states, "this lesion is 87% likely to be malignant," the radiologist needs to understand which features prompted that assessment. Without that transparency, it remains a black box making life-or-death decisions, and no one is comfortable with that.

Parallel Testing

Run the AI-powered solution alongside current methods to compare outputs directly. This allows buyers to:

- Observe where the AI solution performs differently.
- Understand error patterns and accuracy improvements.
- Build confidence through direct comparison.
- Establish baseline performance in their specific environment.

One radiologist, initially skeptical, revised his viewpoint mid-meeting when the AI flagged a subtle anomaly they had missed—remarking, "I wouldn't have caught that on a busy shift." That moment shifted the discussion. Suddenly, the focus was not on replacing radiologists but on augmenting their capabilities during high-volume periods when fatigue becomes a factor.

For Maya's imaging system, this meant processing the same diagnostic images through both their existing system and the new AI-enhanced solution, then comparing results with confirmed diagnoses.

Continuous Validation Framework

This goes beyond simply achieving "yes"—it's about ensuring your solution continues to perform as promised long after the sale closes. Buyers seek proof that validation is not a one-time act but a sustained commitment.

Establish processes for ongoing validation post-implementation, including:

- Regular performance reviews against established benchmarks.
- Mechanisms for human override when necessary.
- Continuous learning and improvement protocols.
- Transparency in system updates and changes.

This ongoing validation builds confidence that the solution will continue to perform effectively over time, not just during the initial implementation.

AI-Assisted Validation Tools

While validating AI-powered solutions presents unique challenges, AI can enhance the validation process itself when applied appropriately.

Risk Assessment Algorithms

AI-powered risk assessment tools can analyze the buyer's environment and your solution to identify potential implementation challenges that might be overlooked in manual reviews. These tools:

- Compare the current environment to previous implementations.
- Identify risk factors based on established patterns.
- Quantify the potential impact of various risks.
- Suggest mitigation strategies based on past successes.

For healthcare technology like Maya's imaging system, risk assessment algorithms might evaluate factors such as IT infrastructure compatibility, staff readiness, workflow integration points, and regulatory compliance requirements.

Alignment Verification Tools

AI can assess the fit between your solution and the buyer's stated requirements with greater precision than manual methods. These tools:

- Map solution capabilities to specific buyer requirements.
- Identify gaps or misalignments requiring attention.
- Suggest configuration adjustments to improve fit.
- Highlight areas needing additional validation.

For complex medical equipment, alignment verification might analyze hundreds of technical specifications against the hospital's requirements, identifying potential issues that would be challenging to spot manually.

Reference Matching Engines

These tools aid in matching prospects with past customers who have faced similar challenges, making the conversation feel less like a scripted pitch and more like seasoned advice from someone who has been there.

AI can identify the most relevant references for a specific buyer by analyzing attributes and implementation patterns. These tools:

- Match prospective buyers with existing customers based on multiple attributes.
- Identify references with similar challenges or constraints.
- Suggest specific aspects of the reference implementation most relevant to the prospect.

- Prepare customized discussion guides for reference conversations.

For Maya, a reference matching engine might identify which existing hospital customers had similar patient demographics, technical infrastructure, and implementation challenges to the prospect.

Using AI Ethically in Validation

When leveraging AI for validation support, maintain ethical standards by:

Transparency: Accuracy: Complementary
Data: The goal is to leverage AI to enhance validation, making it more thorough and relevant while preserving the human relationships that foster trust in complex sales.

From Validation to Rehearsal

Effective validation naturally leads to the next step in the Apex 8 methodology: Rehearsal. The insights gained during validation become essential inputs for preparing your formal presentation.

Use your validation results to:

Incorporate Prepare Build This groundwork ensures that your presentation does more than pitch a solution—it confirms one the buyer already believes in, supported by data they helped validate.

For Maya, the validation process with the hospital uncovered specific concerns about radiologist workflow and training requirements. This allowed her to refine her approach and prepare a presentation that directly addressed these issues, supported by evidence from the validation activities.

A solution is ready for formal presentation when:

- Key technical, business, and human fit questions have been validated.
- Significant risks have been identified, and mitigation strategies have been developed.
- The buyer has experienced enough of the solution to feel confident in the outcomes.
- Both you and the buyer have a clear understanding of implementation requirements.
- The validation process itself has built relationships and trust with key stakeholders.

The speed and scrutiny of today's sales environment demand more than insight—they require co-created clarity. A well-crafted validation process does not merely describe the future; it allows the buyer to experience it firsthand. That's your edge.

Applying Validation Principles in Practice

Let's conclude with practical applications of these validation principles in your next client interaction:

Design Validation That Matters

This is your moment to show—not just tell—that your solution fits perfectly. Your validation plan isn't merely a checklist—it's your strongest proof of insight.

For your next opportunity:

1. Identify the top three concerns or uncertainties the buyer has about your solution.
2. Design a specific validation activity for each concern.
3. Involve the buyer directly in planning the validation approach.
4. Create clear success criteria that both parties agree represent meaningful validation.
5. Document validation results in a format that can be shared with stakeholders not directly involved.

Address All Three Dimensions

Return to the three dimensions—technical, business, and human—from earlier in this chapter, and ensure your plan includes clear validation steps for each.

When developing your validation plan:

1. Include at least one validation activity for each dimension.
2. Match validation methods to the aspects of greatest concern for key decision-makers.
3. Consider the evidence standards expected in the buyer's industry and organization.
4. Include validation of AI components if your solution utilizes intelligent systems.
5. Create a timeline that allows for thorough validation without unnecessarily extending the sales cycle.

Leverage References Strategically

As discussed earlier in the Reference Alignment Strategy, specificity matters. Here's how to take it further—combine technical, financial, and end-user perspectives into a holistic reference experience.

To enhance validation through references:

1. Identify reference customers based on relevant attributes, not just their willingness to advocate.
2. Prepare reference conversations focused on specific validation questions.
3. Include technical, business, and end-user references when possible.
4. Follow up reference discussions with a synthesis of key validation points.
5. Consider bringing prospect and reference teams together for direct dialogue.

Use AI Appropriately

To enhance your validation with AI:

1. Use AI tools to identify potential implementation risks based on the buyer's specific situation.
2. Generate customized validation plans based on the buyer's industry, size, and requirements.
3. Analyze previous implementations to identify the most relevant validation approaches.
4. Maintain human oversight of all AI-generated validation materials.
5. Focus your time on the relationship and trust-building aspects of validation that AI cannot replicate.

Remember, effective validation is not about checking boxes or adhering to a rigid process. It's about creating experiences that transform skepticism into confidence—for both the buyer and you as the seller. When done well, validation not only increases the likelihood of winning the deal but also lays the foundation for successful implementation and a lasting customer relationship.

In the next chapter, we will explore the fourth step in the Apex 8 methodology: Rehearsal. We will examine how to prepare yourself and your team to present your validated solution with maximum impact.

As you move forward, maintain the discovery mindset from Chapter 2, the collaborative approach to solution development from Chapter 3, and the evidence-based validation approach we've explored here. The combination of these elements creates a foundation for technical sales success that neither traditional approaches nor AI alone can match.

CHAPTER 5

REHEARSE: PREPARING FOR IMPACT IN THE COGNITIVE COMPUTING ERA

The Performance Gap

Alex stood in the empty conference room, digital presentation materials loaded and robotics simulation models prepared. Tomorrow, he would present an $8.2 million automated assembly solution to the executive team at Peterson Manufacturing, a leading automotive components manufacturer. After weeks of discovery, collaborative prototype development, and thorough validation with their engineering team, this was the moment of truth—the formal presentation to their decision-makers.

As the Technical Solutions Director for Nexus Robotics, Alex had a decade of experience designing and selling complex industrial automation systems. He knew the technology inside and out. The solution for Peterson was technically sound, financially justified, and thoroughly validated. By all rational measures, he should feel confident.

Yet something nagged at him. Peterson's CEO was known for asking penetrating questions about workforce impact. Their COO had a reputation for probing deeply into implementation timelines. The CFO would certainly demand detailed payback scenarios. And the VP of Operations—a former line worker who had risen through the ranks—would be skeptical about any promises of seamless integration.

Alex had the answers to all these potential questions somewhere in his materials. But could he deliver them with the clarity, confidence, and conviction needed in the moment? Could he navigate the complex group dynamics that would unfold? Could he handle unexpected objections or last-minute concerns?

The difference between winning and losing this opportunity wouldn't be the solution itself; it would be his ability to present it effectively when it mattered most.

This is the rehearsal challenge in technical selling—the fourth step in the Apex 8 methodology.

In the AI era, where buyers expect extraordinary levels of preparation and polish, the gap between adequate and exceptional presentations has widened dramatically. AI has raised the bar for visual quality, data presentation, and technical narrative. Simultaneously, the human elements of connection, conviction, and authentic expertise have become more valuable precisely because they cannot be easily replicated by artificial intelligence.

Effective rehearsal bridges this gap, transforming technical expertise into compelling performance when it matters most.

Beyond Traditional Preparation

In today's selling environment, presentations are no longer about proving what you know—they're about shaping what the buyer believes.

Traditional preparation for technical sales presentations typically involves:

1. Creating slides and visual materials
2. Reviewing technical specifications and details
3. Perhaps a quick run-through of the presentation flow

This level of preparation was sufficient when presentations were primarily about information transfer—communicating technical capabilities, specifications, and proof points to an audience that lacked access to this information.

But in the AI era, where buyers have already researched your solution extensively and can access detailed technical information instantly, the purpose of presentations has fundamentally shifted. Presentations are no longer primarily about information transfer; they are about:

- Creating a shared vision that aligns stakeholders
- Building confidence in implementation success
- Demonstrating a deep understanding of the buyer's specific situation
- Creating an emotional connection to outcomes and value
- Navigating complex group dynamics and decision processes

This shift requires a fundamentally different approach to preparation—one that treats the presentation as a high-stakes performance requiring deliberate rehearsal across multiple dimensions.

The consciously competent technical seller understands the distinction between practice and rehearsal:

- **Practice** focuses on familiarity with material—running through your presentation to know what comes next.

- **Rehearsal** focuses on performance excellence—systematically refining how you deliver, respond, and adapt in the moment.

For Alex at Nexus Robotics, true rehearsal meant not just reviewing his slides or practicing his talk track but systematically preparing for the full performance demands of presenting a complex automation solution to Peterson Manufacturing's diverse stakeholders.

The Three Dimensions of Effective Rehearsal

Effective rehearsal in technical selling addresses three essential dimensions of performance. Neglecting any dimension creates a vulnerability that can undermine even the most technically sound solution.

Content Rehearsal

Content rehearsal focuses on mastering what you'll communicate—your technical narrative, value story, and supporting evidence. It addresses questions like:

- Is our core message clear, compelling, and consistent?
- Have we structured information for maximum impact and retention?
- Are complex technical concepts translated into accessible language?
- Does our narrative connect technical capabilities to business outcomes?
- Have we prepared multiple levels of detail for different audience needs?

For Alex's robotics presentation, content rehearsal meant refining how he would explain the automated assembly system's technical architecture, integration approach, and implementation methodology in terms that resonated with each stakeholder's primary concerns.

Content rehearsal techniques include:

- **Story**: Crafting and testing the overall narrative flow
- **Message**: Identifying and sharpening 3-5 core messages
- **Technical**: Converting complex concepts into accessible language
- **Layered**: Organizing supporting information in accessible layers
- **Value**: Explicitly linking technical capabilities to business outcomes

The key to content rehearsal is not just knowing your material but structuring and expressing it for maximum clarity and impact with your specific audience.

Delivery Rehearsal

Delivery rehearsal focuses on how you'll communicate—your verbal and non-verbal expression, visual support, and overall presence. It explores questions like:

- Does our delivery style match the audience's expectations and culture?
- Are we using language that resonates with different stakeholder groups?
- How will we handle the physical environment and technology?

- Do our visual and demonstration elements support rather than distract?
- Does our delivery convey appropriate confidence and expertise?

For the industrial automation presentation, delivery rehearsal included preparing for the factory environment where ambient noise might affect audibility, ensuring the robotics simulation would display properly on the available screens, and practicing transitions between technical explanation and value discussion.

Delivery rehearsal techniques include:

- **Environment**: Rehearsing in conditions similar to the actual presentation
- **Technology**: Testing all demonstration elements and visuals
- **Verbal**: Adjusting pace, tone, and terminology for the audience
- **Non-verbal**: Ensuring body language supports your message
- **Energy**: Matching your presentation energy to the setting and culture

Effective delivery rehearsal creates a foundation of confidence that allows you to focus on connection and adaptation during the actual presentation rather than worrying about basic performance elements.

Interaction Rehearsal

Interaction rehearsal focuses on preparing for the dynamic elements of presentation—questions, objections, stakeholder reactions, and group dynamics. It addresses challenges like:

- How will we handle anticipated questions and objections?

- What if stakeholders disagree with each other during the presentation?
- How should we respond to unexpected challenges or concerns?
- What if the energy or direction shifts during the presentation?
- How will multiple presenters coordinate and support each other?

This phase draws heavily on insights from Discovery and Validation—because the questions you prepare for should reflect what stakeholders truly care about, not just what you think they'll ask.

For Alex's robotics presentation, interaction rehearsal meant preparing for the CEO's workforce impact concerns, the COO's timeline skepticism, and the potential tension between engineering enthusiasm and operations caution.

Interaction rehearsal techniques include:

- **Objection**: Identifying likely challenges and preparing responses
- **Stakeholder**: Predicting reactions and interactions among decision-makers
- **Question**: Practicing answers to difficult or technical questions
- **Dynamic**: Developing strategies for unexpected turns
- **Team**: Defining roles and handoff protocols for multiple presenters

Effective interaction rehearsal transforms potential challenges from threats into opportunities to demonstrate expertise and build confidence.

Rehearsal Methods for Complex Solutions

Different technical sales situations require different rehearsal approaches. The consciously competent seller selects methods that address the specific challenges of their presentation context.

Story Arc Development

Best for: Complex solutions with multiple components or stakeholders.

Story arc development creates a coherent narrative structure that guides the entire presentation, connecting technical elements to business outcomes through a compelling journey. This approach:

- Establishes a clear beginning, middle, and end to the presentation
- Creates natural transitions between technical and business aspects
- Builds toward key moments of insight or decision
- Maintains focus on outcomes rather than features

For Alex's robotics presentation, story arc development meant structuring his presentation as a journey from current manufacturing challenges through implementation to future operational excellence, with the automated assembly system serving as the enabler of transformation rather than the focus itself.

Effective story arcs for technical presentations often follow patterns such as:

- Challenge → Solution → Transformation
- Problem → Approach → Outcomes
- Current State → Implementation Journey → Future State

The power of story arc development lies in its ability to create coherence across complex technical information, making the overall narrative accessible even when individual components are sophisticated.

Technical Narrative Simplification

Best for: Highly technical solutions requiring explanation to mixed audiences

Technical narrative simplification transforms complex technical concepts into accessible explanations without sacrificing accuracy. This technique:

- Identifies core technical principles that non-specialists need to understand
- Creates progressive layers of detail appropriate for different stakeholders
- Develops analogies and visual models that illuminate key concepts
- Builds vocabulary bridges between technical and business language

For example, to explain digital twinning, Alex described it as "a real-time mirror of the production line that lets us test changes virtually before applying them in reality," making the concept instantly relatable to non-technical stakeholders. Instead of diving into the data structures and IoT architecture behind it, he focused on the practical impact: "It means you can validate process changes without disrupting actual production and see exactly how modifications will affect throughput before committing a single dollar."

The art of technical narrative simplification is maintaining technical integrity while creating accessibility. The goal is not to oversimplify but to create multiple pathways to understanding based on the audience's background and interests.

Objection Anticipation and Response

Objections aren't hurdles—they're your greatest opportunity to demonstrate empathy, insight, and adaptability in real time.

Best for: Solutions with significant change impact or inherent concerns

Objection anticipation systematically prepares for the concerns, questions, and challenges likely to arise during a presentation. This approach:

- Identifies potential objections from different stakeholders
- Develops evidence-based responses for each concern
- Prepares for both rational and emotional objections
- Creates natural transitions from objection to solution advantage

For the robotics presentation, objection anticipation meant preparing for workforce concerns from the CEO, timeline skepticism from the COO, and integration doubts from the VP of Operations—not just with answers, but with evidence and examples that would resonate with each stakeholder.

Effective objection response follows patterns such as:

- Acknowledge → Bridge → Reframe
- Validate → Evidence → Advantage
- Listen → Understand → Resolve

The goal is not to "overcome" objections but to transform them into opportunities to demonstrate understanding and build confidence.

Stakeholder Reaction Mapping

Best for: Presentations to multiple decision-makers with different priorities

Stakeholder reaction mapping predicts how different audience members will respond to various aspects of your presentation and prepares adaptive strategies. This technique:

- Analyzes each stakeholder's priorities, concerns, and communication style
- Identifies potential points of alignment and tension between stakeholders
- Prepares tailored explanations and evidence for different perspectives
- Develops strategies for navigating group dynamics

For Peterson Manufacturing, stakeholder reaction mapping meant understanding the relationships and potential tensions between the executive team members, particularly between the efficiency-focused COO and the workforce-concerned CEO.

The power of stakeholder reaction mapping lies in its ability to transform unpredictable group dynamics into anticipatable patterns that you can prepare to navigate effectively.

Environment and Technology Preparation

Best for: Presentations with significant demonstration elements or challenging settings

Environment and technology preparation ensures that physical and technical factors support rather than undermine your presentation. This approach:

- Simulates actual presentation conditions as closely as possible
- Tests all technology elements under realistic circumstances
- Prepares contingency plans for technical or environmental challenges
- Optimizes setup for both presenter and audience experience

For industrial robotics, this meant ensuring the simulation software would run properly on available equipment, preparing backup access methods, and accounting for potential factory floor conditions such as ambient noise or lighting challenges.

Effective environment preparation eliminates distractions and technical issues that could undermine an otherwise excellent presentation. It creates a foundation of confidence that allows you to focus on audience connection rather than technological concerns.

Team Role Definition and Coordination

Best for: Complex presentations requiring multiple presenters

Team role definition creates clarity around responsibilities and coordination for multi-person presentations. This approach:

- Assigns clear roles based on expertise and relationship with the audience
- Establishes seamless transitions between presenters
- Creates protocols for supporting each other during questions

- Ensures consistent messaging across different presentation segments
- Aligns every team member on key messages, transitions, and non-verbal cues that create a seamless, supportive delivery

For the robotics presentation, this might involve coordination between Alex as the solution architect, an implementation specialist for timeline questions, and perhaps a financial analyst for detailed ROI discussions.

Effective team coordination makes the presentation feel cohesive rather than fragmented, with each presenter enhancing rather than competing with others.

The AI Presentation Revolution

These rehearsal strategies prepare you to meet a new challenge: the elevated expectations set by AI-enhanced content and hyper-informed buyers.

The emergence of AI has transformed both the creation and evaluation of technical presentations in ways that demand new approaches to rehearsal.

Rising Presentation Expectations

AI-powered presentation tools have dramatically raised the bar for visual quality, data presentation, and narrative cohesion. Buyers now expect:

- Professional-grade visuals and animations
- Data visualization that illuminates rather than merely reports
- Seamless transitions between concepts and topics
- Customization that reflects their specific situation

- Flawless technical demonstrations

These expectations create both challenges and opportunities for technical sellers. The challenge is meeting these higher standards; the opportunity is using them to differentiate your approach.

The Authenticity Imperative

As AI-generated content becomes ubiquitous, the value of authentic human expertise, judgment, and connection has increased. Buyers are increasingly sensitive to presentations that feel scripted, generic, or over-rehearsed.

This creates what I call the "presentation paradox"—the need to be simultaneously more polished and more authentic than ever before. You resolve this paradox not by choosing one over the other, but by rehearsing so thoroughly that you're free to focus on the room—not your notes. True polish is what lets authenticity shine through.

Effective rehearsal navigates this paradox by creating a foundation of confidence that allows for genuine presence and connection during the actual presentation.

Balancing Technology and Humanity

The most effective technical presentations in the AI era balance technological sophistication with human connection. They leverage AI tools to enhance rather than replace human expertise, creating presentations that are:

- Visually compelling without feeling artificial
- Data-rich without losing narrative coherence
- Technically precise without becoming inaccessible

- Thoroughly prepared without feeling scripted
- Responsive to audience dynamics while maintaining direction

For Alex's robotics presentation, this meant using sophisticated simulation tools to visualize the automated assembly system while ensuring his own expertise and understanding of Peterson Manufacturing's specific challenges remained central to the presentation.

AI-Augmented Rehearsal Tools

While we've focused on the challenges AI creates for presenters, AI tools can also significantly enhance rehearsal effectiveness when properly applied.

Presentation Analysis Systems

AI-powered analysis tools can evaluate presentation content and structure, identifying potential improvements that might be overlooked in manual reviews. These tools:

- Assess narrative flow and logical progression
- Identify jargon or complex terminology that may create barriers
- Evaluate the balance between technical detail and business outcomes
- Suggest structural improvements for clarity and impact

In one case, a presentation analysis tool flagged that 70% of a robotics pitch focused on technical specifications, prompting the team to reframe it around business impact, which ultimately swayed the CFO. That single insight transformed their presentation from "here's our amazing technology" to "here's how this technology transforms your profitability."

For technical presentations, these tools help ensure that sophisticated content remains accessible to diverse stakeholders without sacrificing accuracy or depth.

Delivery Coaching Tools

AI coaching systems can provide feedback on verbal and non-verbal delivery elements that impact presentation effectiveness. These tools analyze:

- Vocal patterns, including pace, tone, and emphasis
- Language choices, including clarity and accessibility
- Energy levels and engagement indicators
- Non-verbal elements captured through video practice

For Alex's robotics presentation, a delivery coaching tool might identify when his explanation of machine learning algorithms became too technical or when his energy dropped during the ROI discussion.

Objection Simulation Engines

AI can help prepare for questions and challenges by simulating the types of objections likely to arise in specific sales contexts. These tools:

- Generate realistic questions based on solution type and industry
- Create stakeholder-specific challenges based on roles and priorities

- Provide frameworks for effective responses
- Allow for practice of difficult interactions

Think of it like a sparring partner who's read your buyer's mind—it throws punches so you can learn to dodge gracefully. Instead of being caught off guard by a brutally direct question about ROI timelines, you've already practiced a dozen variations of the answer.

For industrial automation, an objection simulation engine might generate the workforce impact questions the CEO is likely to ask, allowing Alex to refine his responses before the actual presentation.

Using AI Ethically in Rehearsal

When leveraging AI for rehearsal support, maintain ethical standards by:

1. **Be:** Let AI enhance your strengths, not manufacture someone you're not.
2. **Balance**: Combine AI analysis with human feedback.
3. **Stay**: Use AI to refine key areas—not the entire performance.
4. **Trust**: AI offers patterns, not your buyer's unique context.

The goal is to use AI to enhance your preparation while maintaining the authentic expertise and connection that differentiate human sellers from automated alternatives.

Once you've integrated these AI-supported insights into your performance plan, you're ready to move from preparation to the main stage.

From Rehearsal to Presentation

Effective rehearsal naturally leads to the next step in the Apex 8 methodology: Present. As you complete your rehearsal process, several final preparations ensure you're ready for the actual performance:

Mental Preparation

Beyond technical readiness, prepare your mental state for optimal performance:

- Visualize success, including handling challenging moments effectively
- Develop pre-presentation rituals that center and focus your energy
- Identify specific points where you'll check in with your audience
- Set realistic expectations that allow for adaptation
- Create separation between rehearsal and performance mindsets

Rehearsal is your training ground; performance is game day. Don't try to control every play—trust the practice, read the room, and respond in real time.

For Alex, mental preparation included visualizing confident responses to the CEO's workforce questions and the COO's timeline concerns, preparing himself to view these moments as opportunities rather than threats.

Environmental Verification

Revisit your earlier environment and tech setup plans—now test them under real-world conditions:

- Arrive early to test all systems under actual conditions
- Verify seating arrangements and sightlines for optimal engagement
- Check sound levels and visibility from different positions
- Ensure all demonstration elements are functioning properly
- Prepare backup options for any critical technical components

This verification eliminates preventable distractions and creates confidence that the environment will support your performance.

Team Alignment

For multi-person presentations, ensure final coordination:

- Confirm roles and responsibilities for each segment
- Review protocols for handling questions and transitions
- Establish non-verbal signals for timing or adaptation needs
- Align on key messages and priorities
- Create pre-presentation connections among team members

This alignment ensures that the presentation feels cohesive rather than fragmented, with each presenter enhancing the overall impact.

A presentation is ready when:

- Core content has been refined and structured for maximum impact

- Delivery has been practiced under conditions similar to the actual presentation
- Likely questions and objections have been anticipated and prepared for
- Technology and environment have been verified, and contingencies established
- Mental and emotional readiness has been established

Applying Rehearsal Principles in Practice

Let's conclude with practical applications of these rehearsal principles for your next important presentation:

Develop a Rehearsal Strategy

For your next opportunity:

1. Assess which rehearsal dimension needs the most attention: content, delivery, or interaction
2. Allocate preparation time proportionally to address your specific challenges
3. Create a structured rehearsal plan rather than generic practice time
4. Identify specific aspects of your presentation to refine through deliberate rehearsal
5. Use recording and feedback to make rehearsal more effective

Create Realistic Conditions

To enhance your rehearsal effectiveness:

1. Simulate the actual presentation environment as closely as possible—see 'Environmental and Technology Preparation' for setup best practices

2. Practice with the same technology you'll use in the final presentation
3. Enlist colleagues to play the roles of specific stakeholders
4. Introduce realistic interruptions and questions during practice
5. Vary conditions to prepare for different possible scenarios

Prepare for Key Moments

To build confidence for critical presentation elements:

1. Identify the 3-5 most important moments in your presentation
2. Rehearse these segments with extra attention and repetition
3. Prepare multiple approaches to critical explanations or responses
4. Create seamless transitions into and out of these key moments
5. Develop recovery strategies if these moments don't go as planned

These moments often align with the narrative peaks defined during your story arc—rehearse them with intentionality to maximize emotional and strategic impact.

Leverage AI Appropriately

To enhance your rehearsal with AI:

1. Use AI tools to analyze your presentation content for clarity and structure
2. Record practice sessions for AI-assisted delivery analysis
3. Generate potential objections and questions with AI simulation tools

4. Create visual and data elements that meet AI-era expectations
5. Balance AI-powered polish with authentic presence and expertise

Remember, effective rehearsal isn't about memorization or scripting. It's about creating a foundation of confidence and capability that allows you to be fully present and responsive during the actual presentation. The goal is preparation that enables authentic performance rather than mechanical recitation.

If you've implemented the strategies outlined above—structured rehearsal, simulated conditions, stakeholder preparation—you're ready. Well-executed rehearsal ensures your presentation lands with clarity and confidence.

In the next chapter, we'll explore the fifth step in the Apex 8 methodology: Present. We'll examine how to deliver your solution with maximum impact, building on the discovery, prototyping, validation, and rehearsal work you've completed.

But as you move forward, recognize that strong rehearsal is what transforms preparation into performance. The technical seller who masters rehearsal evolves from a knowledgeable expert into a compelling advocate for the buyer's future success.

CHAPTER 6

PRESENT: DELIVERING IMPACT IN THE ARTIFICIAL INTELLIGENCE LANDSCAPE

The Meta-Moment

As we reach this chapter on presentation—the fifth step in the Apex 8 methodology—it's worth pausing to acknowledge a meta-moment: you've been experiencing the methodology even as you've been learning about it.

Throughout this book, we've been following the very framework we're teaching:

- In Chapter 1, we conducted **Discovery** of your current reality as a technical seller in the AI era, the challenges you face, and your vision for success.
- In Chapters 2-3, we developed a **Prototype** of the Apex 8 methodology, outlining its components and inviting your mental collaboration in adapting it to your situation.
- In Chapter 4, we explored approaches to **Validate** these concepts through evidence and examples relevant to your selling environment.
- In Chapter 5, we prepared you to **Rehearse** the methodology's application in your specific context.

Now, we arrive at **Present**—the moment when the full value of what we've built together becomes clear and compelling. This deliberate structure wasn't accidental; it demonstrates the natural flow of the methodology itself, progressing from understanding to solution, to verification, to preparation, and finally to presentation.

But what exactly is a meta-moment? It's when you transcend the immediate situation to recognize larger patterns and possibilities. A meta-moment occurs when you step outside the frame of the current activity to see it from a higher perspective—understanding not just what you're doing, but the broader implications and opportunities it creates.

In technical selling, meta-moments are particularly powerful. They are the instances when you realize you're not just selling a product but redefining an entire category. When you're not merely responding to stated requirements but reshaping how the customer fundamentally thinks about their challenges. When you're establishing new standards that will become industry benchmarks or creating language that will define future conversations.

This is your moment—not to impress, but to ignite change. You're not just presenting a product; you're enabling a better future they can see, feel, and believe in.

This meta-structure reveals a fundamental truth about effective technical selling: presentation isn't a standalone event but the culmination of a carefully orchestrated process. When you follow the Apex 8 methodology, presentation becomes not the beginning of persuasion but its fulfillment—the moment when all previous work converges to create impact.

And here's perhaps the ultimate meta-moment of modern selling: the very concept we're discussing—asymmetric knowledge—has completely flipped. Thirty years ago, sellers held a knowledge advantage that gave them power in sales relationships. Today, buyers often arrive with deeper research, more comprehensive analysis, and broader context than any individual salesperson could possess. The asymmetry hasn't disappeared; it's reversed. This isn't a problem to lament but a transformation to embrace. When knowledge is no longer your advantage, relationships become your differentiator. When information is abundant, wisdom becomes rare. When analysis is automated, empathy becomes essential.

This is why presentation in the Apex 8 methodology isn't about proving what you know—it's about demonstrating how you think, how you connect, and how you can guide complex decisions with insight that transcends mere information. In an era of knowledge symmetry, your value isn't what you've memorized, but what you understand about transforming knowledge into action.

During the presentation itself, you create meta-moments for your customers—instances when they suddenly see beyond the immediate features and functions to recognize how your solution might transform their entire approach to a problem. These are the moments that separate transactional selling from truly transformative technical sales.

The Trust Culmination

Jennifer stood before the executive team of Meridian Utilities, an energy provider serving three states. As Chief Solutions Architect for GridSync Technologies, she had spent months guiding Meridian through the evaluation of a comprehensive smart grid solution—an $18 million investment that would transform their aging infrastructure into an AI-optimized, resilient energy network.

The stakes were enormous. Blackouts during recent weather emergencies had created public backlash. Renewable integration was straining existing systems. Regulatory pressures were mounting. Yet the implementation would touch every aspect of their operations, from distribution automation to customer engagement.

Jennifer didn't begin with a standard capabilities presentation. Instead, she displayed a map of Meridian's service territory with weather patterns superimposed, showing vulnerable sections of their grid.

"Over the past two months, we've worked together to understand these vulnerabilities," she began, referencing their discovery process. "Your team helped us design this solution architecture," she continued, acknowledging their collaborative prototyping. "And last week, we validated the approach through simulation testing with your engineers and operations team," she added, connecting to their validation activities.

This wasn't a vendor pitch. It was a shared milestone in a co-created journey—evidence of trust built through discovery, design, and validation.

This is the power of the Apex 8 approach to presentation—building on established trust rather than attempting to create it from scratch during a single meeting.

Beyond Information Transfer: The Art of Emulation

Modern presentation is not about relaying facts—it's about emulation: demonstrating the customer's reality, challenges, and opportunities as if they're already solved, in real time, through your solution.

Traditional technical presentations focus primarily on information transfer—communicating capabilities, specifications, and potential value to an audience presumed to be starting from limited knowledge. This approach has become increasingly obsolete in the AI era, where buyers often arrive with extensive information already in hand.

Modern technical presentation serves a fundamentally different purpose—one best understood as emulation rather than information transfer. In this context, emulation means actively mirroring and solving the customer's specific processes, challenges, and opportunities through your technology, right before their eyes.

When you truly emulate the customer's world through your presentation:

Painting Them Into the Picture

The classic sales maxim to "paint the customer into the picture" takes on new meaning in technical selling. Through emulation, you're not just describing how they might use your solution; you're showing them their own processes, challenges, and opportunities reflected and improved through your technology.

This is the difference between telling someone about a solution and letting them see themselves successfully using it. In psychological terms, you're creating recognition and ownership before any purchase decision.

For Jennifer's smart grid presentation, this meant showing actual Meridian distribution networks operating under simulated storm conditions, demonstrating how their specific infrastructure would respond with GridSync's technology in place. The executives weren't watching a generic demonstration; they were seeing their own grid, their own challenges, and their own success.

Creating Experiential Recognition

Effective emulation creates moments of recognition where customers think, "That's exactly our situation" or "That's precisely what we're trying to accomplish." These recognition moments build credibility and connection far more powerfully than abstract claims or generic examples.

The consciously competent presenter deliberately engineers these recognition moments, selecting examples, scenarios, and visualizations that precisely mirror the customer's reality. This isn't manipulation; it's relevant customization that demonstrates a genuine understanding of their situation.

Solving Problems in Real-Time

Perhaps most powerfully, emulation involves solving the customer's specific problems in real time during the presentation itself. Rather than talking about theoretical capabilities, you demonstrate actual problem resolution using their data, their scenarios, and their success criteria.

This approach transforms the presentation from a passive experience to an active demonstration of value creation. The customer doesn't have to imagine how your solution might work in their environment; they see it working before their eyes.

The AI Advantage in Emulation

The AI era has dramatically enhanced our ability to create true emulation in technical presentations. Unlike traditional customization, which prepares tailored content in advance, AI-powered emulation responds in real time—mirroring the customer's world with live data and adaptive interfaces.

Technologies that would have required months of custom development can now be configured in days or even hours. Visualizations that would have been impossibly complex can now be generated dynamically. Customer-specific scenarios that would have required extensive programming can now be simulated on the fly.

This capability creates a fundamental shift in what's possible during technical presentations, allowing for levels of customization and demonstration that were previously infeasible. The consciously competent seller leverages these capabilities not as technical showcases but as emulation tools that make the abstract concrete and the potential immediate.

The Three Realities Framework in Action

To guide this emulation process, the Three Realities framework introduced earlier becomes your structural anchor—turning recognition into resonance, and resonance into momentum.

In Chapter 2, we introduced the Three Realities framework for discovery: Current Reality, Problem Reality, and Vision Reality.

This same framework provides a powerful structure for technical presentations, creating a natural narrative flow that connects with diverse stakeholders.

Current Reality: Establishing Common Ground

The presentation begins with Current Reality—a shared understanding of the buyer's present situation. This isn't just a perfunctory acknowledgment; it's a deliberate demonstration of the depth of your understanding.

For Jennifer's smart grid presentation, Current Reality included:

- A detailed map of Meridian's existing infrastructure with highlighted vulnerability points
- Operational metrics showing current performance against industry benchmarks
- Regulatory compliance challenges documented through their audit history
- Customer satisfaction data correlated with outage patterns

This Current Reality section creates three crucial effects:

1. **Validation**: It confirms your understanding of the buyer's specific situation, demonstrating that you've truly listened during discovery.
2. **Relevance**: It establishes that your presentation is customized to their environment, not just a generic solution pitch.
3. **Trust**: It reminds stakeholders of the work you've already done together, strengthening the foundation of the relationship.

The key to an effective Current Reality presentation is specificity—using the buyer's actual data, terminology, and metrics rather than general industry observations.

Problem Reality: Creating Tension and Urgency

From Current Reality, the presentation moves to Problem Reality—the challenges, risks, and missed opportunities created by the current situation. This section generates productive tension that motivates change.

For the utility presentation, Problem Reality included:

- Predictive models showing vulnerability to future weather events
- Competitive analysis revealing market share erosion to more resilient providers
- Regulatory risk assessment forecasting potential penalties
- Customer impact narratives connecting outages to real-world consequences

Problem Reality produces several essential effects:

1. **Urgency**: It makes the status quo uncomfortable, creating motivation to act rather than defer.
2. **Scope**: It defines the full dimensions of the challenge, often broader than initially perceived.
3. **Shared**: It ensures all stakeholders recognize the same problems, creating alignment on the need for change.

The art of Problem Reality presentation is balance—creating enough tension to motivate action without overwhelming or demoralizing the audience.

Vision Reality: Inspiring Commitment

The presentation culminates with Vision Reality—a compelling portrayal of the future state enabled by your solution. This isn't just about features and capabilities; it's about transformation and outcomes.

For the smart grid presentation, Vision Reality included:

- Resilience modeling showing outage reduction during simulated weather events
- Customer experience scenarios demonstrating new capabilities for end users
- Operational dashboards revealing unprecedented visibility and control
- Regulatory compliance projections showing future-proofed compliance

Effective Vision Reality creates powerful effects:

1. **Emotional**: It helps stakeholders feel the benefits, not just understand them intellectually.
2. **Ownership**: It reflects the buyer's input from earlier collaborative steps, making it their vision, not just yours.
3. **Momentum**: It creates forward motion toward implementation rather than further evaluation.

The key to an effective Vision Reality presentation is making it both aspirational and achievable—a future that inspires while feeling realistic and attainable.

The Art of Technical Storytelling

Presenting complex technical solutions requires more than just organizing information—it demands true storytelling that makes sophisticated concepts accessible without oversimplification. The consciously competent technical seller masters several storytelling techniques:

Contrast and Transformation

As demonstrated in Vision Reality, transformation stories resonate strongest when they show real contrast. This technique reinforces that emotional arc—making progress feel vivid, not theoretical.

Effective technical stories use contrast to highlight value and differentiation:

- Juxtaposes before and after states to make benefits tangible
- Highlights critical differences between approaches
- Creates memorable comparisons that demonstrate impact

For the smart grid presentation, Jennifer might contrast storm response under the current system (36-hour outage restoration) versus the new solution (predictive rerouting preventing most outages entirely), making abstract capabilities concrete through specific scenarios.

Technical Translation Through Analogy

Analogies and metaphors translate complex technical concepts into accessible understanding:

- Connect unfamiliar technologies to familiar experiences
- Bridge technical and business language

- Create "mental hooks" for retaining complex ideas

For utility executives unfamiliar with machine learning, Jennifer might compare the grid optimization system to an experienced air traffic controller who learns patterns over time, anticipates problems before they occur, and makes rapid adjustments to maintain flow.

Progressive Disclosure

Once analogies open the door, progressive disclosure helps guide different stakeholders through increasing layers of complexity— without losing anyone along the way.

Technical storytelling requires careful sequencing of information:

- Begin with concepts accessible to all stakeholders
- Layer in complexity progressively as understanding builds
- Create "on-ramps" for different technical levels

Instead of overwhelming with technical details upfront, Jennifer might start with outcome-focused descriptions of the smart grid solution and progressively introduce more technical details as stakeholder understanding and interest develop.

Narrative Bridges

Connecting technical capabilities to business outcomes requires explicit narrative bridges:

- "This technical capability enables this operational improvement..."
- "This system feature translates to this business advantage..."
- "This architectural approach solves this strategic challenge..."

These bridges ensure that technical features aren't presented in isolation but are directly connected to the outcomes stakeholders care about.

Data-Driven Narrative

Technical storytelling balances narrative and data:

- Uses data to establish credibility and specificity
- Embeds quantitative evidence within qualitative narrative
- Creates visual representations of data that reveal patterns and insights

For example, Jennifer's animated outage chart demonstrated a dramatic drop in grid disruption after implementing predictive re-routing—instantly reframing the solution from theoretical to transformative.

For the utility presentation, Jennifer might use animated data visualization to show how predictive maintenance algorithms reduced outages in similar implementations, making the statistical case within a compelling narrative framework.

Positioning: The Strategic Context of Your Solution

Beyond the mechanics of presentation, the consciously competent technical seller understands the power of positioning—the strategic context in which your solution is perceived. Positioning shapes how buyers evaluate your offering relative to alternatives, including the status quo.

The Trust Transfer Principle

Just as validation established credibility through buyer-specific modeling and peer alignment, trust transfer reinforces that credibility by anchoring it in familiar, respected reference points.

Effective positioning leverages what I call the "trust transfer principle"—associating your solution with entities, concepts, or approaches the buyer already trusts. This creates a powerful psychological shortcut to credibility.

The most sophisticated technical sellers deliberately position their offerings alongside trusted reference points:

- **Category**: Placing your solution within a recognized and respected category ("This is enterprise-grade infrastructure automation...")
- **Technology**: Connecting to trusted technology standards or platforms ("Built on the same secure foundation used by leading financial institutions...")
- **Conceptual**: Framing your approach within established and respected methodologies ("This follows the same principles as high-reliability organizations in aerospace...")
- **Brand**: Creating associations with trusted brands through partnerships, integrations, or comparisons

We see this positioning strategy at work throughout the technology industry. When enterprise technology companies sponsor Formula One racing teams, they're not just seeking visibility—they're creating psychological associations between their solutions and the precision engineering, performance, and reliability that Formula One represents. When a buyer who trusts and admires the engineering excellence of these racing teams sees Oracle, SAP, or HP logos on the cars, that trust transfers, however subtly, to the technology providers.

In Jennifer's utility presentation, she might position GridSync's smart grid solution alongside the reliability systems used in air traffic control or reference its foundation in the same fail-safe architecture used in nuclear power safety systems—creating associations with domains where reliability is paramount and trusted.

Effective positioning isn't deceptive; it's a legitimate way to help buyers contextualize complex offerings within frameworks they already understand and trust. The key is to ensure that the associations are genuine and substantive, not merely superficial.

Dynamic Presentation in the AI Era

Beyond content and structure, modern technical presentations demand dynamic delivery that adapts to audience responses in real-time. This adaptive approach:

The AI Acceleration Reality

Before diving deeper into presentation dynamics, let's address what might seem like an overwhelming set of requirements. If you're thinking, "This is an impossible amount of preparation," you're applying pre-AI thinking to an AI-augmented world.

You've already explored how AI supports discovery, prototyping, and rehearsal—now it becomes your partner in real-time adaptation and responsive storytelling.

The reality is that everything we're discussing becomes remarkably manageable when you leverage AI as your presentation co-pilot. By 2025, basic AI literacy won't be optional for technical sellers—

it's as fundamental as email was twenty years ago. With even elementary AI skills, you can:

- Generate multiple versions of your presentation (5-minute, 15-minute, and 30-minute) in seconds.
- Create adaptive modules that can be resequenced on demand.
- Develop response frameworks for potential interruptions.
- Prepare alternative explanations at varying technical depths.
- Generate visualization variants for different audience compositions.

For instance, when Jennifer prepared for her utility presentation, her AI sales assistant analyzed the planned content and automatically generated condensed versions, interruption response protocols, and module variations based on stakeholder interests identified during discovery. This preparation—which would have taken days in the pre-AI era—required only minutes of guidance and review.

Throughout this book, we assume you have at least fundamental AI literacy. You don't need to be a prompt engineering expert, but you should be using purpose-built sales AI tools that understand presentation dynamics. Without this basic capability, you're attempting to compete in the modern sales environment with a significant disadvantage—like trying to stream HD video over dial-up.

The approaches we're discussing aren't complex when AI-augmented; they're simply the new standard for professional presentation in technical sales. What seems like excessive preparation becomes routine efficiency when you're properly equipped with the tools of the era.

Reading and Responding to Stakeholder Signals

Beyond individual cues, you also need to track how stakeholders interact with each other—surfacing tensions, alliances, or silent disagreements that may influence final decisions.

Effective presenters continuously monitor verbal and non-verbal cues from their audience:

- Engagement indicators like leaning forward, note-taking, or nodding.
- Confusion signals such as furrowed brows or side conversations.
- Interest patterns revealed by questions or increased attention.
- Skepticism shown through crossed arms, leaning back, or questioning looks.

For Jennifer's utility presentation, noticing the COO's increased attention during discussions on reliability might prompt her to expand on that aspect, while recognizing the CFO's skeptical expression during ROI projections could signal the need for additional evidence.

Time Compression and Interruption Management

A critical reality of presentations in the AI era is the compression of available time and the increased frequency of interruptions. Today's executives have unprecedented demands on their attention, and AI-empowered stakeholders often arrive with specific questions that may redirect your planned flow.

You've already prepared modular content during rehearsal. Now, your job is to deploy it intuitively—responding to the room with poise and precision.

The consciously competent presenter prepares for these realities by:

1. **Priority**: Identifying the absolute must-communicate elements versus those that are supplementary.
2. **Interruption**: Developing techniques for gracefully handling interruptions while maintaining narrative coherence.
3. **Demonstration**: Creating demonstration components that can stand alone or be integrated into the broader presentation as time allows.

At another pitch, a presenter stuck rigidly to a 40-minute flow when the CTO unexpectedly asked for "just the core in 10." The result? Rushed delivery, missed priorities, and a confused buying team.

Again, this might sound like a daunting amount of preparation, but with AI support, these variations can be generated and refined in minutes rather than hours. Your AI sales assistant can instantly create condensed versions of presentations, suggest priority hierarchies based on stakeholder profiles, or develop interruption response frameworks tailored to your content.

For Jennifer's utility presentation, this might mean having her AI assistant prepare a condensed version of her smart grid benefits that she can deploy when the CEO unexpectedly has only 15 minutes instead of the scheduled hour, or having ready access to specific technical details when an executive interrupts with targeted questions.

This flexibility isn't about abandoning structure but about creating adaptive frameworks that maintain impact regardless of time constraints or flow disruptions. It represents another dimension of trust-building—demonstrating respect for the buyer's time and priorities through your ability to adapt without losing effectiveness.

Moving Between Conscious Competence and "The Zone"

The best presenters move fluidly between conscious technique and intuitive flow—what performers call "the zone." This balance allows you to switch between structure and spontaneity depending on the moment.

The most effective technical presenters develop a fluid ability to move between conscious competence and unconscious competence—this dynamic shifting between deliberate technique and natural flow is particularly crucial during presentations.

When you're "in the zone" during a technical presentation:

- Your responses to questions flow naturally from deep expertise rather than rehearsed scripts.
- You read the room intuitively, sensing engagement and skepticism without conscious analysis.
- Your authentic passion for the solution emerges naturally.
- You adapt seamlessly to unexpected turns without apparent effort.
- Technical concepts translate into accessible explanations without deliberate calculation.

This state of unconscious competence is particularly powerful during key presentation moments:

- When addressing complex technical concepts that require natural fluency.

- During objection handling where authentic responses build trust.
- In moments of unexpected challenge where scripted answers would feel wooden.
- During demonstrations where technical expertise must appear effortless.
- In closing sequences where conviction matters more than precise wording.

When you feel the room shift—leaning in, nodding, following your lead—and you know you're in sync, that's when presentation becomes presence.

The ability to enter this zone doesn't happen by accident—it's built on the foundation of thorough preparation and simulation. The preparation work you did during the rehearsal phase creates the confidence that allows you to trust your unconscious competence when it matters most.

For Jennifer's utility presentation, this meant preparing so thoroughly that when the CFO interrupted with an unexpected question about integration costs, she could smoothly transition into the zone—responding with authentic expertise rather than panicked recollection of prepared answers. Her unconscious competence allowed her to maintain connection while delivering precise information, something no purely scripted approach could achieve.

The rhythm between conscious and unconscious competence is crucial—knowing when to deliberately apply techniques and when to trust your expertise to flow naturally. In general:

- Use conscious competence when setting up structured elements like frameworks or technical explanations.

- Transition to unconscious competence during moments requiring authentic connection or adaptation.

- Return to conscious competence when summarizing key points or establishing next steps.

Learning to trust yourself enough to enter the zone during high-stakes presentations is one of the most powerful skills in technical selling. It requires both thorough preparation and the courage to let go of conscious control at key moments, allowing your authentic expertise to emerge naturally.

Real-Time Adaptation Techniques

The art of adaptation in technical presentations requires continuous tuning:

- Adjusting technical depth based on comprehension signals.

- Expanding or contracting examples based on time constraints.

- Resequencing content modules to follow emerging interests.

- Shifting emphasis between technical and business benefits as needed.

Managing Group Dynamics

Technical presentations often involve multiple stakeholders with different priorities and perspectives. Effective presenters:

- Identify and address emerging tensions between stakeholders.

- Create balanced attention across different decision-makers.

- Facilitate productive dialogue rather than allowing derailment.
- Use disagreement as an opportunity to clarify and address concerns.

For the utility presentation, it is essential to acknowledge the tension between the operations team's focus on reliability and the finance team's concern about cost. Then, demonstrate how the proposed solution addresses both priorities.

Visual and Experiential Elements

The AI era has dramatically raised expectations for visual and experiential elements in technical presentations. Buyers now expect sophisticated visualizations that illuminate rather than merely accompany verbal content.

Beyond Slides: Multisensory Presentation

Effective technical presentations engage multiple senses:

- Visual elements that reveal patterns and relationships
- Interactive components that create hands-on understanding
- Spatial arrangements that clarify complex systems
- Audio elements that enhance comprehension or emotional connection

For the smart grid presentation, this might include a physical model of a substation with an augmented reality overlay showing energy flow patterns, or an immersive audio experience of control room communications during an outage response.

Data Visualization for Technical Clarity

Complex data becomes accessible through thoughtful visualization:

- Comparative visualizations that highlight differences between options
- Progressive visualizations that show changes over time
- Relationship visualizations that reveal connections between elements
- Interactive visualizations that respond to "what if" scenarios

Jennifer might use an interactive map allowing executives to trigger simulated weather events to see how the smart grid would respond compared to their current infrastructure—making abstract capabilities tangible through visual experience.

While these tools can dramatically elevate clarity and engagement, top sellers are increasingly realizing that even the best AI-generated content often requires a human touch to truly resonate.

The Human Touch: AI's Final Mile Problem

As with other steps in the Apex 8 methodology, AI provides a powerful foundation—but not the full solution. While AI has revolutionized presentation development, a crucial insight for the consciously competent seller is what I call the "Final Mile Problem."

Modern sales platforms built around the Apex 8 methodology provide tremendous value in automating and enhancing presentation development. They help create compelling visuals, data representations, and demonstration elements that are far more sophisticated than possible without AI assistance. However, the most successful technical sellers recognize when to incorporate human expertise for that crucial final portion.

Particularly with graphics, design, and creative elements, AI lacks the wisdom, insight, and anticipation of human reactions that make truly exceptional presentations resonate. AI cannot fully predict how a specific human audience will emotionally respond to visual elements, nor can it perfectly anticipate unspoken concerns that might arise during a presentation. This human element—the ability to sense, adapt, and connect on an intuitive level—remains beyond AI's capabilities in 2025.

This may mean having a designer refine key visuals for critical presentations, collaborating with a creative director on essential demonstration elements, or bringing in specialized expertise for particularly sophisticated visualization needs.

This isn't a failure of AI but simply the reality of where the technology stands. The consciously competent seller knows when AI augmentation is sufficient and when human expertise should be engaged for maximum impact.

Demonstration Approaches for Confidence Building

Demonstrations have evolved beyond mere feature showcases to become confidence-building experiences:

- Outcome-focused demonstrations that show results, not just features
- Stakeholder-directed explorations that follow audience interests
- Stress-test scenarios that demonstrate resilience under challenging conditions

- Comparative demonstrations that illustrate before-and-after or with-and-without contrasts

Rather than simply demonstrating monitoring capabilities, Jennifer might show a simulated storm response, comparing current manual processes with the automated response of the smart grid solution—demonstrating not just features but meaningful outcomes.

What demonstration techniques could you use in your next presentation to help your buyer visualize not just features—but future results?

Virtual and Augmented Reality for Complex Solutions

For particularly complex solutions, virtual and augmented reality create understanding that is impossible through traditional methods:

- Immersive environments that allow stakeholders to "experience" solutions
- Scale transformations that make massive infrastructure comprehensible
- Physical/digital overlays that connect abstract systems to concrete reality
- Time compression that shows long-term effects in compressed timeframes

For instance, one utility company used a VR simulation to let executives walk through a digitally reconstructed disaster zone, illustrating how the proposed smart grid would have rerouted power and minimized outage time—converting abstract ROI metrics into a visceral experience.

For utility infrastructure spanning thousands of square miles, virtual reality might allow executives to "fly through" their grid, identifying vulnerability points and experiencing how the new solution would address them—creating understanding that would be impossible through slides alone.

From Presentation to Commitment

The presentation culminates in the critical transition to commitment—setting up the next step in the buying process. This transition isn't abrupt but flows naturally from the presentation itself.

Reading Buying Signals

Consciously competent presenters recognize signals indicating readiness for commitment:

- Forward-looking questions about implementation or timing
- Stakeholder conversations about internal next steps
- Specific inquiries about contracts or terms
- Expressions of ownership or future-oriented language

These signals often emerge during the presentation, creating natural transition opportunities.

Natural Transition Techniques

Instead of awkward closing techniques, effective transitions flow organically:

- Summarizing points of agreement established during the presentation
- Acknowledging remaining questions and proposing specific resolution paths

- Suggesting logical next steps based on what's been presented
- Framing the decision as a continuation of progress already made

For Jennifer, this might involve noting, "Based on what we've covered today, it seems the main remaining question is implementation phasing. We've prepared three potential approaches that we could review next Tuesday with your operations team."

Setting Up Finalization

The presentation sets up the sixth step of Apex 8: Finalize. This setup includes:

- Confirming understanding of the buyer's decision process
- Establishing specific next actions and owners
- Addressing any immediate concerns that could impede progress
- Creating momentum that carries forward into closing

The goal is not to force an immediate decision but to create a clear and natural path toward commitment.

Applying Presentation Principles in Practice

These practical strategies aren't just about making your presentations visually appealing—they're about reinforcing trust, sharpening business impact, and guiding buyers to confident decisions.

Let's conclude with practical applications of these presentation principles in your next client interaction:

Structure for Maximum Impact

For your next presentation:

1. Organize content using the Three Realities framework (Current, Problem, Vision)
2. Begin by reinforcing your understanding of the buyer's specific situation
3. Create deliberate contrast between current challenges and future possibilities
4. Position your solution within contexts already trusted by the buyer
5. Connect every technical capability explicitly to business outcomes

Enhance Visual Experience

To elevate your visual presentation:

1. Replace at least one text-heavy slide with a visual representation
2. Create before-and-after visualizations that make benefits tangible
3. Develop one interactive element that allows stakeholder exploration
4. Use consistent visual language that reinforces your positioning
5. Design slides as visual support, not as a presentation script

Prepare for Dynamic Delivery

To enhance your presentation adaptability:

1. Identify potential stakeholder reactions and prepare adaptive responses

2. Create modular content that can be resequenced based on audience interest
3. Prepare alternative examples for different stakeholder priorities
4. Practice transitioning between different levels of technical depth
5. Develop natural language for checking understanding and adjusting approach

Build on Trust Foundation

To leverage the trust established in earlier steps:

1. Reference specific insights gained during discovery and prototyping
2. Acknowledge stakeholder contributions to the solution approach
3. Validate evidence and examples with stakeholders who participated in earlier steps
4. Use language that reinforces partnership rather than vendor-client separation
5. Create natural transitions to commitment that build on established progress

Remember, effective presentation in technical selling isn't solely about performance or persuasion. It's about realizing the trust and understanding you've built through discovery, prototyping, validation, and rehearsal. When these foundations are established, presentation becomes not a high-pressure selling event, but the natural culmination of a collaborative journey.

In the next chapter, we'll explore the sixth step in the Apex 8 methodology: Finalize. We'll examine how to convert the momentum generated during your presentation into clear commitment and preparation for successful implementation.

As you move forward, recognize that a strong presentation transforms preparation into momentum. The technical seller who masters presentation evolves from a trusted advisor into a catalyst for action—helping buyers move confidently from consideration to commitment.

CHAPTER 7

FINALIZE: FROM CONSIDERATION TO COMMITMENT WITH AI

Enhancement Beyond Closing

Dr. Nadine Chen looked across the conference table at the research director of Hamilton Biosciences. After three months of careful work—discovery conversations with their scientific team, collaborative prototype development of a custom mass spectrometry solution, thorough validation through sample testing, meticulous rehearsal, and a compelling presentation to their leadership—they had reached the moment of decision.

The $1.2 million advanced research system would transform Hamilton's proteomics capabilities, potentially accelerating their drug discovery timeline by months. The ROI analysis showed a clear path to value. Technical validation confirmed compatibility with their workflows. The scientists were enthusiastic. Everything logically pointed to a "yes."

Yet Nadine recognized the subtle hesitation in Dr. Raymond's body language—a slight tension, a barely perceptible pulling back.

"The proposal looks comprehensive," he said, tapping the document. "We just need to take some time to think it through and discuss it internally before making a commitment of this magnitude."

In a previous era, Nadine might have launched into objection-handling mode—asking what specific concerns needed addressing, offering additional evidence, or applying subtle pressure tactics to secure the deal before leaving. But as the Solution Architect for Precision Analytics, she understood a fundamental truth about technical selling in the AI era:

When you've built genuine trust through the Apex 8 process, hesitation rarely stems from unaddressed objections. It almost always comes from fear.

"Dr. Raymond," she said gently, "we've worked closely with your team for three months now. Your scientists helped design this solution. We've validated the approach with your actual samples. Is there something specific about the solution that doesn't feel right, or is it more about the natural caution that comes with any significant investment?"

The question created a moment of insight. "It's not the solution," he admitted. "Four years ago, we invested in a system from another vendor that never delivered as promised. The board hasn't forgotten, and neither have I. We need to be absolutely certain this time."

This wasn't an objection to overcome—it was a fear pattern to help him move beyond using the trust they'd built together.

"That makes perfect sense," Nadine nodded. "Previous experiences shape our caution. Would it help if I outlined the specific differences in our implementation approach that address exactly what went wrong last time? And perhaps we could structure the agreement with performance milestones that give you verification points before full commitment?"

The tension in Dr. Raymond's face eased. "That would actually be very helpful."

This is finalization in the Apex 8 methodology—not applying pressure or manipulative closing techniques, but leveraging established trust to help buyers move past fear toward a decision that serves their genuine interests.

In the AI era, where information is abundant and verification is instant, traditional closing tactics have become not just ineffective but counterproductive. The consciously competent technical seller recognizes that the path to commitment isn't through persuasive pressure but through trust leverage that creates a safe passage to "yes."

The New Decision Psychology

Traditional sales methodologies treat finalization—what they typically call "closing"—as a distinct phase requiring specific techniques to overcome resistance and secure commitment. This approach emerged in an era when salespeople held significant information advantages over buyers and could use that asymmetry to create pressure.

The AI era has fundamentally transformed this dynamic in several crucial ways:

The Transparency Revolution

Today's buyers can instantly verify claims, compare alternatives, and access expert opinions about technical solutions. AI-powered analysis tools help them evaluate proposals against industry benchmarks, identify potential implementation risks, and simulate outcomes with unprecedented sophistication.

This transparency means that misrepresentation, exaggeration, or pressure tactics don't just fail—they create lasting damage to trust that spreads quickly through professional networks and digital channels. The consciously competent seller recognizes that transparency isn't a threat but an ally that validates genuine value.

The Trust Acceleration Effect

As we explored in Chapter 3, the Apex 8 methodology creates accelerated trust development through collaborative discovery, prototyping, validation, rehearsal, and presentation. This trust foundation fundamentally changes the psychology of the commitment moment.

In traditional sales, closing often feels adversarial—the seller trying to "get" commitment from a reluctant buyer. In Apex 8, finalization becomes a natural extension of the collaborative journey already undertaken. The question shifts from "should we buy this?" to "how do we move forward together?"

The Fear Reality Triangle

Despite all the logical analysis and verification tools available, technical buying decisions still carry emotional weight, particularly fear. The consciously competent seller recognizes that at the moment of commitment, three specific fears often emerge even when all logical objections have been addressed:

1. **Fear**: Previous negative experiences with similar solutions or vendors create caution that transcends the current situation.

2. **Fear**: Complex technical implementations always contain elements that can't be fully predicted, creating anxiety about potential complications.

3. **Fear**: Decision-makers worry about how they'll be judged if the implementation encounters challenges, potentially affecting their professional reputation.

These fears aren't irrational or unreasonable—they're normal human responses to significant decisions with uncertain outcomes. The finalization approach in Apex 8 doesn't dismiss these fears or attempt to override them with pressure; it acknowledges them and leverages trust to help buyers move through them confidently.

The Trust Leverage Framework

Effective finalization in technical selling requires a framework for leveraging established trust to move past hesitation toward confident commitment. This framework has four components:

Recognition: Identifying Fear Patterns

The first step in leveraging trust is recognizing when hesitation stems from fear rather than unresolved questions or legitimate concerns. Fear patterns typically manifest in several recognizable ways:

- **Sudden**: The conversation shifts from specific solution details to broad, generalized concerns about implementation, cost, or risk.
- **Regression**: The buyer falls back on formal procurement or approval processes as a reason for delay, even when those processes haven't been emphasized previously.
- **Past**: Stories or references to previous negative experiences emerge late in the discussion.

- **Timing**: The projected timeline for decision suddenly extends without clear justification for the additional time needed.
- **New**: Previously unmentioned decision-makers are suddenly introduced as necessary approvers.

In the AI era, these patterns become significantly easier to identify. Modern sales AI tools can analyze conversation patterns, detect shifts in language, and flag potential fear indicators in real time. What once required years of experience to recognize intuitively now becomes visible through AI-enhanced conversation analysis, allowing even newer technical sellers to respond with the wisdom of veterans.

While AI can surface these shifts in tone or phrasing, it's your human judgment that gives them meaning. Pattern recognition is automated—but empathy is earned.

For Dr. Raymond at Hamilton Biosciences, his reference to needing to "think it through," combined with the later revelation of a previous negative experience, clearly indicated a fear pattern rather than an unaddressed objection—a pattern that could be flagged and analyzed by AI to suggest specific trust leverage approaches.

The Continuous Trust Building Reality

It's crucial to recognize that trust building isn't confined to any single step of the Apex 8 methodology—it's a continuous process that accumulates through each interaction. By the time you reach finalization, you've created a substantial trust foundation through:

- **Discovery**: Demonstrating understanding of their specific situation and challenges.
- **Prototype**: Showing competence through collaborative solution development.
- **Validation**: Creating confidence through evidence and verification.
- **Rehearsal**: Preparing thoroughly to respect their time and attention.
- **Presentation**: Delivering clear, honest, and compelling insights that resonate with their reality.

Each of these trust layers contributes to your ability to help buyers move past fear during finalization. When hesitation emerges, you can draw on specific moments from previous steps that demonstrated your competence, character, and commitment to their success.

For example, if Dr. Raymond expresses concern about implementation complexity, Nadine might reference how their collaborative prototype development already addressed integration challenges or how the validation testing confirmed compatibility with their workflows. These specific trust references carry far more weight than generic reassurances.

Articulation: Making Fear Visible

Once you recognize fear patterns, the next step is to help the buyer articulate the specific fear—bringing it into the open so it can be addressed directly. This isn't manipulation; it's a service that aids them in processing their own hesitation.

Effective articulation techniques include:

- **Normalization**: "Many organizations in your position feel cautious about major technology investments given the complexity involved. Is that part of what you're feeling?"
- **Experience**: "Have you had experiences with similar implementations that influence how you're thinking about this decision?"
- **Future**: "If you imagine having made this decision, what concerns come to mind about how it might unfold?"
- **Stakeholder**: "How do you think other stakeholders in your organization view this decision in terms of risk?"

These approaches build on what you already know: fear isn't weakness—it's risk awareness. Your role is to give it shape so it can be handled, not avoided.

Once a fear is articulated, it doesn't disappear—it becomes something that needs to be honored. How you respond to that moment determines whether trust deepens or dissolves.

By bringing fears into the open, you transform them from vague anxieties into specific concerns that can be addressed.

Confirmation: Validating Legitimate Concerns

Once fears are articulated, it is crucial to validate rather than dismiss them. This validation builds on the trust established through previous steps by demonstrating empathy and understanding.

Effective validation approaches include:

- **Explicit**: "That's a completely understandable concern given what happened with your previous implementation."

- **Shared**: "I would be thinking exactly the same thing in your position. That experience would make anyone more cautious."
- **Appreciation**: "Thank you for sharing that concern. It helps me understand your perspective better and ensures we can address it properly."
- **Reframing**: "So it's not about whether our solution would work technically, but about ensuring you don't repeat the negative experience from four years ago."

This validation creates a foundation for moving forward by demonstrating that you take their concerns seriously rather than treating them as obstacles to be overcome.

Resolution: Creating Safe Passage

The final step is creating a path forward that specifically addresses the articulated fears, leveraging the trust you've built to help them move toward commitment with confidence.

Effective resolution approaches include:

- **Differentiation**: Explicitly outlining how your approach differs from past negative experiences, with specific emphasis on elements that address their particular concerns.
- **Risk**: Creating structural elements in the agreement or implementation plan that directly reduce the specific risks they fear.
- **Incremental**: Breaking the decision into smaller steps with verification points that build confidence progressively.
- **Shared**: Establishing clear, mutual responsibilities and success metrics that create confidence in the path forward.

For Nadine with Hamilton Biosciences, this meant offering to structure the agreement with performance milestones that provided verification before full commitment—directly addressing Dr. Raymond's fear of repeating a previous implementation failure.

The power of this framework comes from its basis in trust rather than technique. It honors the buyer's legitimate concerns while helping them move past fear toward a decision that serves their genuine interests.

When Genuine Issues Emerge

While many hesitations stem from fear patterns, the consciously competent seller recognizes that sometimes legitimate, unaddressed issues emerge during finalization. In these cases, the approach shifts from fear resolution to a recursive application of earlier Apex 8 steps.

Distinguishing Issues from Fears

How do you know when you're facing a genuine unresolved issue rather than a fear pattern? Look for these indicators:

- **Specificity**: The concern relates to specific, concrete aspects of the solution rather than general anxiety about outcomes.
- **Novelty**: The issue involves elements not previously discussed or considered during earlier steps.
- **Technical**: The questions focus on particular capabilities, implementations, or outcomes rather than process or emotional concerns.
- **Stakeholder**: The concerns arise from specific stakeholders whose needs may not have been fully addressed earlier.

For example, if Dr. Raymond had instead said, "We're concerned about the integration with our existing chromatography systems which use proprietary data formats," this would indicate a specific technical issue rather than a fear pattern.

The Recursive Approach

When legitimate issues emerge, the Apex 8 methodology becomes recursive—you return to earlier steps as needed:

1. **Mini-Discovery**: Explore the newly emerged concern with the same depth and curiosity you applied in initial discovery.
2. **Rapid**: Develop a focused solution approach specifically addressing the new concern.
3. **Targeted**: Create rapid verification of your approach to the specific issue.
4. **Brief**: Prepare your explanation and evidence carefully.
5. **Focused**: Present your approach to the specific concern clearly and confidently.

This recursive application maintains momentum while ensuring all legitimate issues are addressed. The key is to approach these emergent concerns with genuine interest rather than frustration at the "delay," recognizing that addressing them strengthens rather than weakens the foundation for commitment.

The Speed Imperative

This isn't just about getting to "yes" faster—it's about protecting the opportunity from drift, distraction, and diminishing stakeholder focus.

While recursively addressing legitimate concerns is necessary, the consciously competent seller recognizes the crucial importance of speed in the modern business environment. Competitors are always advancing, market conditions evolve rapidly, and opportunities have increasingly short windows.

This creates what I call the "ethical speed imperative"—the responsibility to help buyers move efficiently toward beneficial decisions without applying inappropriate pressure. This balance is achieved by:

- **Maintaining:** Addressing concerns promptly rather than allowing extended delays.
- **Timeboxing**: Setting clear timeframes for addressing specific issues.
- **Creating**: Helping buyers establish clear criteria and processes for reaching efficient decisions.
- **Highlighting**: Respectfully illuminating the cost of delayed implementation in terms meaningful to the buyer.

For many buyers, delay isn't relief—it's a return to indecision limbo, where motivation fades and fear regains control.

The distinction between pressure and momentum is crucial. Pressure attempts to force a decision despite concerns; momentum helps remove barriers to a decision the buyer wants to make but fears making. The former damages trust; the latter leverages it.

The AI Verification Reality

As we explored earlier in the "Transparency Revolution," buyers today verify claims in real time. The finalization stage is where this capability becomes most decisive.

A unique aspect of finalization in the AI era is what I call the "real-time verification reality"—the fact that buyers can and do use AI tools to verify claims, assess risks, and validate proposals during the decision process itself.

The Transparency Imperative

This verification capability creates an absolute imperative for transparency and accuracy in all aspects of your solution and claims. Consider these realities:

- Buyers can instantly compare your technical specifications against competitor offerings.
- Implementation timelines can be assessed against industry benchmarks.
- Pricing can be evaluated against market standards in real time.
- Reference claims can be verified through digital channels.
- Even subtle exaggerations can be detected through AI analysis.

The consciously competent seller embraces rather than fears this verification reality, recognizing that it validates genuine value while exposing superficial claims.

Creating Verification Pathways

Beyond accepting verification, the most effective approach is proactively creating verification pathways that build final confidence:

- **Evidence**: Providing structured, verifiable evidence for key claims in formats that facilitate AI analysis.
- **Benchmark**: Offering direct comparisons to industry standards that buyers can independently verify.

- **Reference**: Creating direct access to reference customers through digital channels that allow for independent verification.

- **Implementation**: Providing detailed implementation methodologies that can be assessed against established best practices.

- **Risk**: Proactively identifying potential implementation challenges and your mitigation approaches.

For example, when a biotech client hesitated over ROI claims, the sales team used their AI assistant to generate a comparison of time-to-market acceleration versus industry averages—helping the CFO greenlight the deal the same day.

For Nadine's mass spectrometry solution, this might include providing access to performance data from similar installations, detailed integration specifications, and direct connections to scientific teams at reference sites who could verify their experiences.

The Ethics Acceleration Effect

In a world where screenshots last forever and group chats travel quickly, exaggeration doesn't just backfire—it echoes.

The verification reality creates what I call the "ethics acceleration effect"—the understanding that ethical approaches to technical selling are not only morally right but also pragmatically necessary in an era of instant verification.

Misrepresentation, exaggeration, or pressure tactics don't just fail temporarily; they cause lasting damage that spreads exponentially through professional networks and digital channels. A single verified instance of misrepresentation can undermine years of relationship building—not just with one client but across an entire market.

Conversely, demonstrated transparency and accuracy build trust that similarly accelerates, creating reputation capital that facilitates future opportunities. The consciously competent seller recognizes that in the AI era, the fastest path to consistent success is also the most ethical one.

Finalizing Complex Technical Decisions

Beyond addressing fears and leveraging trust, effective finalization in technical selling requires navigating the operational realities of complex organizational decisions.

Navigating Approval Ecosystems

Most significant technical purchases involve multiple stakeholders and formal approval processes.

The consciously competent seller maps and navigates these ecosystems by:

- **Process**: Understanding the formal and informal approval steps required for commitment.
- **Stakeholder**: Ensuring each decision-maker has the specific information they need in formats that address their particular concerns.
- **Documentation**: Structuring proposals and supporting materials to align with internal approval requirements.
- **Approval**: Providing tools and frameworks that help champions navigate internal processes efficiently.

Smoothly navigating approval systems isn't just about logistics—it reassures the buyer that you understand their world and protects the trust you've worked so hard to build.

For Hamilton Biosciences, this might involve preparing specific materials for their board that directly address the previous negative experience, helping Dr. Raymond make a confident case internally.

The Surprise Negotiator Challenge

One of the most challenging scenarios in technical sales is what I call the "surprise negotiator"—a previously uninvolved stakeholder (often from finance or procurement) who enters at the finalization stage with a mandate to extract concessions. This common pattern typically looks like:

"Everything looks great. We want to move forward. Now I'll step aside and our procurement team will handle the final details."

This moment represents a critical trust test. The new stakeholder has no investment in the relationship you've built, doesn't understand the value established through previous steps, and often measures success by the concessions they extract.

While the negotiator may be new to the table, your champions bring the weight of the trust you've built—your job is to help them carry it forward in terms that resonate with the new gatekeeper.

The consciously competent seller prepares for this scenario by:

1. **Early**: Asking about procurement and financial approval processes during discovery to anticipate who might be involved later.

2. **Value**: Creating comprehensive value justification materials specifically designed for financial stakeholders.
3. **Relationship**: When possible, involving procurement or finance representatives earlier in the process to build relationship continuity.
4. **Trust**: Preparing your champions to articulate value to the negotiator in terms that resonate with their priorities.
5. **Negotiation**: Working with your own finance team to establish parameters that protect value while allowing for appropriate flexibility.

In 2025, AI serves as a powerful ally in handling surprise negotiators. Your AI assistant can instantly generate negotiator-specific value summaries, comparative ROI analyses, and industry-specific pricing justifications. What previously required days of financial analysis and document preparation now happens in minutes through AI-augmented preparation.

For example, when faced with an unexpected procurement specialist, your AI sales assistant can immediately analyze your solution's value metrics against industry benchmarks, generate negotiation parameter recommendations, and provide specific responses to common procurement objections—all tailored to your specific deal.

When faced with a surprise negotiator, the key is maintaining connection to the value and trust established throughout the Apex 8 process rather than allowing the conversation to narrow to price alone. This might involve:

- Requesting a joint meeting with both the surprise negotiator and your primary champion who understands the established value.

- Providing a detailed value summary specifically addressing the negotiator's priorities.

- Focusing discussions on total value of ownership rather than initial investment.

- Being prepared to trade non-monetary concessions rather than simply reducing price.

The negotiation phase is not separate from the trust-building process—it's a continuation where you apply the same principles of understanding, adaptation, and value focus that have guided all previous steps.

The Ethical Crossroad: When Trust Meets Hard Negotiation

One of the harsh realities about the surprise negotiator scenario? It's ethically questionable and can feel like a punch to the gut. You've been negotiating in good faith, and suddenly they spring this on you at the last minute. They know you've invested $30,000 to $50,000 of your company's money, team time, and travel expenses. They know you're too far in to walk away. You're going to feel like you've been duped.

I've been blindsided like this more than once in my career, and I didn't see it coming. Initially, I was furious at the customer—feeling like they'd violated the trust we'd built. But here's where the real difference between human and AI capability becomes stark.

AI will pivot on pure logic. It'll instantly suggest adjustments, alternatives, and negotiation tactics. Yes, by all means, use what you can. In fact, your first move will be to pull out your phone or tablet and say, "AI friend, I've got some bad news for us; we just got hit with some challenging news," and let it run the scenarios to give you options.

But here's what AI can't do: make the human decision about how to proceed. This is literally what they pay you the big bucks for. This is where you earn your money. Tomorrow, you're going to be on the phone with someone who knows nothing about you, and you're going to have to shine.

The interesting dynamic? This new negotiator might have AI too, and if they did their homework, they know everything that happened. But here's the kicker—they might choose not to acknowledge that fact. They might play ruthlessly and pretend they don't know about your investment of time and resources. A naive AI would make the mistake of admitting what it knows, but a human negotiator never will.

This is where you have to step out of the AI suit entirely and become Tony Stark for a minute. You'll have to make assumptions, negotiate hard, and go old school. Because when you're facing a last-minute ethical dilemma, AI becomes your little brother in sales—the greatest gopher ever, the best ideation partner—but only you can decide your next move.

And here's what I've learned: you have the green light to say what needs to be said. "I cannot proceed in this negotiation unless the people I've worked with this entire time are in the room. If you think that 15% you're about to squeeze out is worth it, let me be clear—it's not. This approach is damaging the relationship in ways you haven't considered. I'm not walking away, but if we proceed this way, I cannot and will not assure you there won't be lasting damage to our business relationship. This is an ethical situation we're in now."

When I've brought this ethical reality to light, I've seen backpedaling and sudden reconsiderations. But my point is this: never get lost in AI. It's your incredibly efficient assistant, your dynamic ideator, but it needs you to navigate the ethical gray zones. That's how you know it's truly yours—and that you're still essential to the process.

Addressing Last-Mile Technical Concerns

Technical questions often emerge during final reviews. Effective resolution requires:

- **Response**: Providing thorough, accurate answers with minimal delay.
- **Appropriate**: Calibrating the technical depth of responses to the stakeholder's background and needs.
- **Evidence**: Supporting technical assertions with verifiable evidence.
- **Expert**: Providing direct connections to technical specialists when needed for specific concerns.

The goal is to resolve technical questions definitively rather than partially, eliminating any lingering uncertainty that could delay commitment.

Once last-mile concerns are addressed, the next challenge is often scope—transitioning from interest to implementation in a way that feels safe and manageable.

Building Momentum Through Incremental Commitments

When full commitment isn't immediately possible, creating a structured path of incremental steps helps maintain momentum:

- **Phase**: Establishing formal commitments to specific next steps with clear deliverables.

- **Pilot**: Creating limited-scope deployments that build confidence for full implementation.

- **Staged**: Structuring agreements with clear milestones and verification points.

- **Success-Based**: Defining specific metrics that trigger expansion from initial implementation to broader deployment.

For the mass spectrometry solution, this might involve an initial installation focused on a specific research project, with predetermined criteria for expanding to full deployment based on validated results.

The Power of Collective Confidence

Effective finalization acknowledges that in organizational decisions, individual confidence must transform into collective confidence. This transformation requires:

- **Confidence:** Equipping internal champions with the tools and information they need to build confidence among other stakeholders.

- **Objection:** Preparing champions for potential challenges they might face internally.

For example, when Nadine prepared Dr. Raymond for the board review, she equipped him with technical evidence slides and case studies—ensuring he could handle tough questions about ROI and integration without hesitation.

- **Decision:** Providing clear, evidence-based rationales that supporters can use to explain the decision to others.
- **Implementation:** Creating tangible pictures of post-implementation success that can be shared across the organization.

The most successful technical sales create not just individual buyers but also internal advocates who carry confidence throughout their organization.

From Finalization to Debrief

Effective finalization naturally leads to the seventh step in the Apex 8 methodology: Debrief. As commitment is secured, several actions prepare for this crucial learning step:

Capturing Decision Insights

Each finalization is more than a win or a loss—it's a reflection of your method. What you capture here becomes the blueprint for your next success.

During the finalization process, valuable insights emerge about how the buyer makes decisions, what factors influenced them most, and which aspects of your approach proved most effective. Documenting these insights while fresh creates valuable intelligence for:

- Improving future approaches with similar clients.
- Refining your understanding of decision drivers in this market.

- Enhancing your own conscious competence through reflection.

For the Hamilton Biosciences opportunity, Nadine might note how a previous negative implementation experience influenced their decision process and which specific assurances ultimately built their confidence.

Setting Implementation Expectations

The moments immediately following commitment are crucial for establishing realistic expectations about implementation. This includes:

- Confirming key milestones and timelines.
- Establishing communication protocols for the implementation phase.
- Identifying potential early challenges and how they'll be addressed.
- Setting clear roles and responsibilities for both teams.

This expectation setting creates a foundation for successful implementation and a positive debrief experience.

Building Relationship Continuity

Finalization also establishes the relationship framework that will continue through implementation and beyond:

- Introducing implementation team members who will be involved.
- Creating direct connections between counterparts in both organizations.
- Establishing regular review and communication cadences.

- Defining how success will be measured and celebrated.

These actions transform the sales relationship into an implementation partnership, setting the stage for long-term success.

A commitment is ready for debrief when:

- All stakeholders have confirmed their support.
- Formal agreements are completed or in final processing.
- Implementation expectations are clearly established.
- The relationship foundation for successful delivery is in place.

Applying Finalization Principles in Practice

Let's conclude with practical applications of these finalization principles in your next client interaction:

Prepare for Fear Patterns

Before your next finalization conversation:

1. Research the buyer organization's history with similar implementations.
2. Identify potential fear patterns based on their industry and situation.
3. Prepare specific trust leverage approaches for each potential fear.
4. Create structural elements that could address common concerns.
5. Develop incremental commitment options if full commitment seems unlikely.

Your AI sales assistant makes this preparation remarkably efficient, analyzing client history, industry patterns, and your discovery notes to generate fear pattern predictions and response strategies automatically. What once required days of manual analysis now happens in minutes through AI-augmented preparation.

Create Verification Confidence

You started building verification confidence during your presentation—now, you're turning that credibility into commitment by making independent validation easy.

To enhance buyer confidence through verification:

1. Audit your solution claims for absolute accuracy and supportability.
2. Organize evidence in formats that facilitate independent verification.
3. Prepare detailed comparisons to industry benchmarks and standards.
4. Create direct access paths to reference clients.
5. Develop transparent risk assessments with specific mitigation approaches.

AI verification tools now make this process seamless, automatically generating verification packages, comparative analyses, and risk assessments based on your solution parameters and the client's specific situation. These tools transform what would have been overwhelming preparation into automated processes that enhance your effectiveness.

Accelerate Without Pressuring

To maintain ethical momentum:

1. Make next steps easy to say yes to—and hard to forget.
2. Offer decision support tools, not just deadlines.
3. Help buyers visualize the costs of delay, not just tell them.
4. Be ready with options when momentum slows.
5. Guide your champions through their internal gauntlet.

With AI support, these momentum-building tools can be generated instantly and customized to your specific opportunity. Your AI assistant can create decision frameworks, timeline visualizations, and value-of-time analyses that make the cost of delay tangible without applying unethical pressure.

Leverage Trust Authentically

To make the most of established trust:

1. Reference specific moments from your shared journey that demonstrated value.
2. Connect your solution directly to priorities established during discovery.
3. Acknowledge legitimate concerns while providing confident resolutions.
4. Remind buyers of insights and capabilities demonstrated during validation.
5. Position commitment as the natural next step in your established partnership.

AI-powered conversation intelligence tools can help you identify and track these trust-building moments throughout the sales process, making them easily accessible when needed during finalization. This technology transforms what might seem like complex relationship management into systematic trust leverage.

Remember, effective finalization in technical selling isn't about techniques or tactics. It's about leveraging the trust you've built through discovery, prototyping, validation, rehearsal, and presentation to help buyers move confidently toward decisions that serve their genuine interests.

In the next chapter, we'll explore the seventh step in the Apex 8 methodology: Debrief. We'll examine how to capture crucial learnings from each opportunity to continuously improve your effectiveness as a technical seller.

But as you move forward, recognize that strong finalization is what transforms interest into action. The technical seller who masters finalization evolves from a trusted advisor into a valued partner—helping buyers implement solutions that deliver on the promise of your shared vision.

CHAPTER 8

DEBRIEF: EXTRACTING WISDOM IN A DATA-RICH WORLD

The Continuity of Truth

The conference room was quiet as Rachel pulled up the final slide of her presentation. As the Transportation Solutions Director for ConnectFleet, she had just led her team through a thorough review of their most recent win—a $12 million contract to provide logistics technology and tracking systems for ParcelMaxx, a regional delivery company expanding its operations.

"So we secured the contract," she said, looking around the room at her team. "But that's not why we're here today. We're here to understand exactly why we won, what we could have done better, and how we'll apply these insights to future opportunities and the upcoming implementation."

Six team members sat around the table—the solution architect, sales engineer, pricing specialist, integration consultant, account executive, and product specialist—who had collaboratively sold the complex system. Each had their laptops open, connected to their personal AI assistants to help them capture and process their individual perspectives on the opportunity.

"Before we start, let's remember our rules of engagement," Rachel continued, displaying a slide titled 'Debrief Safety Zone.' "No blame, no shame. We're looking for causes, not culprits. Everyone speaks, everyone's heard. And feedback about process or approach is not a personal attack."

She placed a ridiculous-looking fuzzy hat on the table. "Anyone who violates these principles has to wear the 'Captain Hindsight' hat for five minutes."

The team laughed, tension broken. This wasn't their first debrief using the Apex 8 methodology. They knew the process would be rigorous but constructive—focused on extracting maximum learning rather than on celebration or blame.

"Let's start with the turning point," Rachel prompted. "When did momentum shift decisively in our favor?"

The transportation logistics industry, with its complex technical requirements and multiple stakeholders, creates particularly challenging sales environments. However, the team had found that systematic debriefs—whether after wins or losses—consistently improved their performance across opportunities.

This is the power of the Debrief step in the Apex 8 methodology—the seventh step that transforms experience into wisdom and connects lessons from the past to success in the future.

Beyond Traditional Post-Mortems

Traditional sales reviews typically fall into two problematic categories:

1. **Win:** Cursory reviews of successful deals that focus on congratulations rather than analysis, missing crucial learning opportunities.
2. **Loss:** Blame-oriented examinations of failed deals that create defensiveness and rarely extract useful insights.

The Apex 8 Debrief approach fundamentally differs from both. It recognizes that every opportunity—whether won or lost—contains valuable wisdom that can be systematically extracted, documented, and applied to future success.

What makes this approach transformative is its focus on:

- **Truth:** Maintaining an unbroken chain of factual understanding from what happened to future application.
- **Conscious:** Developing explicit understanding of what worked and what didn't, rather than vague generalizations.
- **Universal:** Applying the same rigorous process to both wins and losses.
- **Forward:** Creating specific, actionable insights rather than just documenting history.

When truth continuity breaks, a win might look like luck, and a loss might look like incompetence—when in reality, both were driven by deeper, repeatable patterns no one stopped to name.

As Michael Jordan famously noted about learning from failure, "I've missed more than 9,000 shots in my career. I've lost almost 300 games. Twenty-six times I've been trusted to take the game-winning shot and missed. I've failed over and over and over again in my life. And that is why I succeed."

The Apex 8 Debrief takes this wisdom further—recognizing that success is equally instructive when properly analyzed. Winning also provides critical lessons, but only if we approach it with the same analytical rigor we apply to losses.

In the ConnectFleet example, Rachel's team won the ParcelMaxx contract, but the debrief would be conducted with the same thoroughness as if they had lost. This discipline ensures continuous improvement rather than complacency—understanding that competitors will study and improve on your success if you don't do it first.

The Psychological Foundation of Effective Debriefs

Before diving into the structure and content of effective debriefs, we must address the psychological foundation that makes them possible. Without this foundation, even the most sophisticated debrief methodology will fail.

Creating Psychological Safety

The single most important element of successful debriefs is psychological safety—the shared belief that team members won't be punished or humiliated for speaking up with ideas, questions, concerns, or mistakes.

Harvard Business School professor Amy Edmondson's research has consistently shown that psychological safety is the key predictor of team learning and performance improvement. In technical sales, where complex solutions and multiple team members create numerous potential failure points, safety becomes even more critical.

The consciously competent sales leader establishes this safety through:

1. **Explicit:** Clearly stated principles that govern the debrief, communicated at the beginning of every session.
2. **Modeling:** Leaders who demonstrate openness about their own mistakes and areas for improvement.
3. **Separating:** Making it clear that critique of actions or decisions is not a judgment of the person.
4. **Equalizing:** Ensuring everyone contributes, not just the dominant voices.

Rachel's "Captain Hindsight" hat provided a playful mechanism for enforcing these principles, but the underlying message was serious: psychological safety is non-negotiable for effective debriefs.

The Play Element in Serious Learning

While debriefs address serious business outcomes, incorporating elements of play significantly enhances their effectiveness. Think of play like oxygen in a fire—it doesn't diminish the burn; it fuels it in just the right way.

The consciously competent seller recognizes that playfulness:

- Reduces defensiveness and ego protection.
- Increases creative thinking and solution generation.
- Creates memorable experiences that reinforce learning.
- Builds team cohesion and trust.

Effective play mechanisms in debriefs include:

- **Playful:** Nerf guns, silly hats, or coin jars for violations of safety principles.
- **Role:** Having team members present from the customer's perspective.
- **Visualization:** Creating physical or visual representations of the sales journey.
- **Metaphor:** Describing aspects of the opportunity through metaphors or analogies.

These elements don't diminish the seriousness of the process but rather enhance it by creating psychological states conducive to honest reflection and creative thinking.

The Apex 8 Debrief Framework

With the psychological foundation established, the Apex 8 methodology provides a structured framework for conducting comprehensive debriefs that extract maximum learning from every opportunity.

I. Mindset & Environment

Every debrief begins with the explicit establishment of the psychological environment:

Rules of Engagement

As introduced earlier, your psychological safety tactics—like "no blame, no shame" and fun accountability rituals—should be clearly posted and modeled to reinforce team norms.

Psychological Safety

- "Fun Penalties" for violations (coin jar, Nerf toss, airhorn).
- Kickoff Ritual (e.g., "One word that describes how you feel about this one").
- Leader models humility first: "Here's what I could've done better..."

For ConnectFleet's debrief, Rachel not only displayed these rules but reinforced them with the playful hat penalty. She might also open by sharing her own perspective on where she could have provided better leadership, setting the tone for honest reflection.

II. The Five Core Prompts

The heart of the Apex 8 Debrief framework consists of five essential questions that guide the team through systematic reflection. These questions apply equally to wins and losses, though the emotional tone might differ:

1. **What point?**

- When did momentum shift—either toward us or away from us?
- What specific moment, conversation, or decision changed the trajectory?
- If we had to identify the fulcrum of this opportunity, what would it be?

2. **Where customer/problem?**

- What assumptions proved incorrect?
- What questions should we have asked but didn't?
- What signals did we miss or misinterpret?

3. **What badly)?**

- Identify specific behaviors, decisions, or work products that had outsized impact.
- What would we definitely repeat in future opportunities?
- What would we absolutely change next time?

4. **What enough?**

- What observations, concerns, or ideas remained unspoken until too late?
- What internal communication barriers affected our approach?
- What customer signals did we notice but not fully address?

5. **What time—specifically?**

- Not generalizations ("communicate better") but specific actionable commitments.
- What concrete changes will we implement in our next similar opportunity?
- Who is responsible for ensuring these changes happen?

For Rachel's team at ConnectFleet, a turning point may have occurred when they demonstrated real-time integration between their tracking system and ParcelMaxx's existing route optimization software—a capability their competitors couldn't match. Their misunderstanding might have stemmed from assuming the CTO was the primary decision-maker, while the COO actually held more influence. What they excelled at was creating a customized ROI model specific to ParcelMaxx's regional expansion plans.

The power of these questions lies in their ability to direct the team toward specific, actionable insights rather than general impressions or feelings about the opportunity.

III. Roles & Leadership

However, powerful questions alone aren't enough; they require the right facilitation, environment, and roles to transform inquiry into insight.

Effective debriefs necessitate clear roles and leadership to maintain focus and psychological safety:

The Facilitator

- Set and maintain the tone (openness, curiosity, levity)
- Call timeouts if discussions become tense or unproductive
- Model vulnerability by sharing personal reflections first
- Protect contributors from judgment or criticism
- Extract and document action items and assign owners
- Maintain a balance between structure and open exploration

Additional Roles May

- **Timekeeper**: Ensures appropriate time allocation across topics
- **Recorder**: Captures key insights and action items
- **Customer:** Consistently represents the customer perspective
- **Devil's:** Constructively challenges consensus thinking

In smaller teams, individuals may take on multiple roles, but the facilitation function remains essential for successful debriefs. AI assistance can significantly enhance the recorder role by providing real-time capture and organization of insights.

IV. AI-Augmented Debriefing

While human leadership anchors the process, AI enhances it—extending what teams can see, remember, and connect over time.

In the AI era, debriefing becomes dramatically more powerful through intelligent augmentation. The consciously competent seller leverages AI in several key ways:

1. **Multi-Layer:**

- An AI notetaker captures the overall session.
- Each team member records individual insights through their personal AI.
- Individual AIs process insights through each person's unique lens.
- Perspectives are shared and synthesized through a master system.
- This creates multiple processing layers that identify patterns a single observer might miss.

2. **Pattern:**

- AI analysis identifies similarities between current opportunities and historical data.
- The system detects recurring challenges or success factors.
- Temporal patterns (e.g., seasonal factors, market cycles) become visible.
- Team performance patterns emerge over multiple opportunities.

For example, an AI assistant might flag that six recent opportunities stalled during legal review—prompting the sales ops team to proactively revise contract templates and accelerate approvals.

3. Insight:

- AI transforms raw discussions into structured insights.
- The system categorizes learnings by relevance to different stakeholders.
- Automatically generates action items from discussions.
- Creates a searchable knowledge base for future reference.

This isn't just summarizing—it's a return to conscious competence. It's how we transform accumulated noise into a clear, layered signal.

4. Continuous:

- AI tracks the implementation of insights from previous debriefs.
- The system identifies which lessons led to measurable improvements.
- Automatically connects past debriefs to current opportunities.
- Continuously refines the debrief process itself.

For ConnectFleet, each team member might use their personal AI to process their individual perspectives on the ParcelMaxx win. The solution architect's AI might identify technical integration factors that proved decisive, while the account executive's AI could surface relationship dynamics that influenced the decision. All these perspectives would then be shared and synthesized into a comprehensive understanding that no single view could provide.

While AI dramatically accelerates and enhances the debrief process, human leadership remains essential for maintaining the psychological foundation and extracting wisdom that machines cannot yet fully comprehend.

Specialized Debrief Approaches

The basic Apex 8 Debrief framework can be adapted to different situations and constraints:

Solo Debriefs

When debriefing solo, the key is curiosity. Can you pause long enough to challenge your assumptions and view the experience from a new angle?

Individual contributors often need to debrief opportunities where they were the primary or only representative. Effective solo debriefing approaches include:

- **AI:** Using conversational AI to ask probing questions and challenge assumptions.
- **Structured:** Following the five core prompts with disciplined self-honesty.
- **Voice:** Recording stream-of-consciousness reflections for AI analysis.
- **Third-Party**: Deliberately adopting the viewpoint of the customer or a colleague.

Speaking your thoughts aloud can unlock clarity that your inner monologue hides. There's power in hearing your own insights form mid-sentence.

The key to effective solo debriefs is creating enough psychological distance to see beyond personal biases and defensive reactions.

Quick Debriefs

When time constraints prevent full debriefs, abbreviated approaches can still extract core learnings:

- **Trigger**: Use just the keywords from the five core prompts outlined earlier to anchor a focused 10-minute session.
- **One-Page**: Complete a standardized form with limited space for each insight area.
- **Time-Boxed**: Set strict time limits (e.g., 15 minutes) with focused facilitation.
- **AI-Accelerated:** Load opportunity data into AI for rapid pattern identification.

For example, under "Turning Point," a team might simply say: "When we mapped their warehouse flow on-screen, they leaned in—we had them from that moment."

These approaches sacrifice some depth for speed but maintain the essential structure that makes debriefs valuable.

Team Flip Reviews

A particularly powerful variant involves team members adopting the customer's perspective:

- Team member(s) role-play as the customer.
- Others interview them about the experience.
- The team reconstructs the opportunity from the customer's viewpoint.

- Insights emerge about how the customer actually experienced the sales process.

This technique reinforces one of the central Apex 8 principles: selling is not what you say—it's what they hear, feel, and experience.

This approach often reveals blind spots in the team's understanding of customer priorities and perceptions.

From Debrief to Handoff

The Debrief step directly feeds into the final step of the Apex 8 methodology: Handoff. This connection transforms debriefs from retrospective exercises into forward-driving mechanisms.

The Universal Handoff Requirement

Every opportunity—whether won or lost—requires a handoff to appropriate stakeholders:

- **For**: Insights and commitments are handed off to implementation/service teams.
- **For**: Learnings are handed off to product teams, leadership, or other stakeholders.

Even when losing a deal, the insights uncovered can inform product strategy, shape competitive positioning, or improve team coordination—making the handoff just as important.

The debrief process builds the foundation for these handoffs, creating structured documentation of what happened, why it happened, and what should happen next.

Building Handoff Documents from Debrief Insights

Effective handoff documents emerging from debriefs typically include:

1. **Opportunity:** Key facts and outcomes.
2. **Critical:** Major learnings from the five core prompts.
3. **Customer**: Deeper understanding of the customer's situation.
4. **Success:** Specific elements that drove the outcome.
5. **Recommendations**: Concrete suggestions for implementation or improvement.
6. **Action:** Specific commitments with owners and timelines.

You don't need a 10-page document—just the right details, for the right people, at the right moment.

The AI-augmented debrief process makes creating these handoff documents significantly more efficient, automatically organizing insights into appropriate categories and formats for different stakeholders.

The Organizational Learning Loop

The Debrief-to-Handoff connection creates a continuous learning loop in organizations that practice the Apex 8 methodology:

1. **Opportunity:** The team extracts insights from the specific opportunity.
2. **Organizational:** Insights flow to appropriate stakeholders.
3. **Systematic:** Changes are implemented based on learnings.
4. **Future:** New approaches are tested in subsequent opportunities.

5. **Effectiveness:** The success of changes is evaluated in future debriefs.

This loop transforms individual debriefs from isolated events into engines of organizational improvement and competitive advantage.

For ConnectFleet, the debrief of the ParcelMaxx win will generate specific handoff documentation for the implementation team, highlighting customer priorities, potential integration challenges, and commitments made during the sales process. It will also provide feedback to product management regarding competitive gaps and feature requests that emerged during the sales cycle.

Commercializing the Debrief: Making Your Wins (and Losses) Everyone's Business

Here's what most companies miss about debriefs: they're sitting on organizational gold and treating it like a sales team exercise. If you want to truly commercialize your organization—to "sell as one," as every executive claims they want—then your debriefs are the missing link.

The Executive Invitation

Sales leaders, this is your moment. Make it psychologically safe for your people to participate, then make it strategic by involving executives. When leadership witnesses a win debrief firsthand, they gain a clear understanding of what resources actually drive revenue. When they sit through a loss debrief, they realize that not every missed deal is the salesperson's fault.

Picture this: Your product manager just heard your reps explain how they lost four votes out of twelve because your accounts payable system requires four screens to clear an AP cycle, while the competitor only needs one. Do you think that will sit on a product roadmap for two quarters? Absolutely not. That's going to be fixed in the next sprint.

This isn't just about keeping leadership informed; it's about equipping them with real-world intelligence that enables smart business decisions. Your CFO might think the travel budget for full-team participation is excessive—until they attend a debrief where the solution architect's on-site expertise made a $3 million difference. Suddenly, that travel line item looks like an investment, not an expense.

The Formula One Principle

Think about Formula One teams. On race day, executives might be busy entertaining sponsors or negotiating next season's contracts. But when it's time for the debrief? Everyone is laser-focused. They understand that this is where months of work are compressed into actionable insights in just a few hours.

Your debriefs operate the same way. A CEO who couldn't attend every client meeting over three months can get caught up on what matters in 90 minutes. But here's the kicker—they need to be there live. A recording or transcript doesn't capture the emotions, conviction, and wisdom that emerge when your team dissects what worked and what didn't. That lived experience is where the real learning occurs.

The AI Era Reality

Here's something sales managers need to understand: in the AI era, the factors that win deals change quarterly. Maybe you've been excelling with technical demos, but suddenly your competitors are using AI to create personalized proof-of-concepts in real time. Your debriefs aren't just learning exercises—they're your early warning system.

Smart sales leaders leverage these sessions as their competitive intelligence hack. When you're in the room listening to your team dissect why they won (or lost), you're gaining real-time market intelligence that no industry report can match. You're learning what works this quarter, not what worked six months ago when someone published a case study.

Making "One Team" Real

Every company talks about breaking down silos. Here's how you actually do it: invite finance to hear why the last deal's price negotiations went the way they did. Bring marketing to understand what messaging resonates in competitive situations. Let product management feel the pain when a feature gap costs you a deal.

When your go-to-market officer sits through a debrief where the rep explains how they pivoted from selling technology to selling business transformation—and that pivot saved the deal—she's not just learning a sales technique. She's receiving a masterclass in how value perception drives revenue.

The key is to make this a regular practice, not a special occasion. Your best and worst deals deserve this attention. Because whether you win or lose, the insights that emerge when the right people are in the room can transform your entire go-to-market strategy.

Applying Debrief Principles in Practice

Let's conclude with practical applications of these debrief principles for your next opportunity:

Prepare Your Debrief Environment

Before your next debrief:

1. Create explicit Rules of Engagement that establish psychological safety.
2. Prepare a physical or virtual space conducive to open discussion.
3. Develop a fun mechanism for maintaining safety norms.
4. Set clear time expectations and an agenda.
5. Ensure all relevant team members can participate.

Leverage AI Effectively

As outlined earlier, AI tools can assist with capture, pattern recognition, and summarization. Your role is to ensure those tools serve human insight—not replace it.

To enhance your debrief with AI:

1. Use a master AI system to capture the overall discussion.
2. Encourage team members to process their perspectives through personal AI assistants.
3. Have AI identify patterns across current and previous opportunities.
4. Generate AI-assisted summaries and action items.

5. Create a searchable knowledge base of insights for future reference.

Connect Debrief to Handoff

To ensure debrief insights drive future success:

1. Identify all stakeholders who should receive handoff documentation.
2. Structure insights in formats relevant to each stakeholder group.
3. Create clear, actionable recommendations for implementation or improvement.
4. Establish accountability for following through on debrief learnings.
5. Build mechanisms to track whether insights led to measurable improvements.

Build Organizational Discipline

To create a culture of effective debriefing:

1. Make debriefs mandatory for all significant opportunities, win or lose.
2. Celebrate insights and improvements, not just sales outcomes.
3. Recognize and reward honest reflection and vulnerability.
4. Share debrief methodologies across teams and departments.
5. Continuously refine your debrief approach based on effectiveness.

Remember, the goal of debriefing isn't to determine who deserves credit or blame. The goal is to extract maximum wisdom from every experience—win or lose—and apply that wisdom to future success. As one experienced sales leader puts it: "Someone is going to learn from your wins and losses. It might as well be you."

In the next chapter, we'll explore the final step in the Apex 8 methodology: Handoff. We'll examine how to transition insights and commitments to implementation teams or other stakeholders, ensuring that the value created through the sales process is fully realized in delivery.

But as you move forward, recognize that effective debriefing is what transforms experience into expertise. The technical seller who masters debriefing evolves from a performer into a learner—constantly improving and adapting through the systematic extraction of wisdom from every opportunity.

CHAPTER 9

HANDOFF: ENSURING VALUE REALIZATION BEYOND THE SALE

When Trust Meets Execution

The conference room at GlobalDefense Corporation was filled with a mix of anticipation and tension. On one side of the table sat Morgan and her team from SecureShield—the sales engineer, solution architect, and account executive—who had spent the last three months guiding GlobalDefense through the selection of an enterprise-wide zero-trust security framework. On the other side sat the implementation specialists responsible for deploying the solution—the project manager, integration lead, and technical specialists who would transform the sold vision into operational reality.

"This isn't just another handoff meeting," Morgan began, opening her tablet. "This is where we ensure that everything we've learned, promised, and validated during the sales process is fully transferred to the implementation team."

The stakes couldn't be higher. GlobalDefense was a defense contractor handling sensitive government data across sixteen countries. The $4.2 million security implementation would impact every aspect of their operations. Any disconnect between what was sold and what was delivered could jeopardize the project and potentially create security vulnerabilities in critical infrastructure.

"We've prepared a dual-layer handoff package," Morgan continued, sharing her screen. "The trust transfer components ensure you understand the relationship dynamics and stakeholder priorities. The technical transfer components provide the architectural specifications and implementation requirements."

She turned to the implementation team. "Before we dive in, I want to be clear: this handoff isn't complete until you have everything you need to succeed. Ask questions. Challenge assumptions. This is your opportunity to gain clarity before you're in front of the client."

The cybersecurity industry, with its complex technical requirements and high stakes, demands exceptionally precise handoffs. But this principle applies across all technical sales: the handoff is where the promises made during sales must connect seamlessly with the reality of implementation.

This is the power of the Handoff step in the Apex 8 methodology—the final step that transforms sale completion into value realization, connecting the entire sales process to successful delivery.

Beyond Traditional Handoffs

Traditional handoffs in technical sales typically fall into two problematic categories:

1. **Administrative:** Cursory sessions focused on contract terms, timelines, and basic specifications, with minimal attention to context or relationship dynamics.

2. **Technical:** Overwhelming information transfers where sales teams "throw the data over the wall" to implementation teams with little structure or prioritization.

Both approaches create dangerous disconnects between what was sold and what is delivered. The result? Implementation teams start from a knowledge deficit, customers experience jarring transitions, and the trust built during the sales process evaporates precisely when it should reinforce confidence in the solution.

The Apex 8 Handoff approach fundamentally differs from both. It recognizes that effective handoffs require a dual-layer transfer that addresses both the trust relationship with the customer and the technical integrity of the solution.

What makes this approach transformative is its focus on:

- **Continuity:** Ensuring the customer feels a seamless transition rather than having to "start over" with the implementation team.
- **Knowledge:** Capturing and transferring not just what was sold, but why decisions were made and what alternatives were considered.
- **Risk:** Explicitly communicating potential implementation challenges identified during the sales process.
- **Relationship:** Transitioning trusted relationships rather than merely technical specifications.

In the SecureShield example, Morgan wasn't just handing off technical specifications for a cybersecurity framework. She was transferring months of relationship building with GlobalDefense stakeholders, knowledge of their security concerns about specific legacy systems, understanding of internal political dynamics between their IT and security teams, and awareness of implementation timing sensitivities related to government contract renewals.

This comprehensive handoff approach ensures that implementation begins with the full context necessary for success, not just a set of technical requirements.

The Dual-Layer Handoff Model

The core of the Apex 8 Handoff approach is the Dual-Layer model—a structured framework that addresses both the human and technical aspects of transition.

Layer 1: Trust Transfer (Emotional/Experiential)

The Trust Transfer layer focuses on the relational, contextual, and experiential aspects of the sale. It ensures that the implementation team understands:

- **Stakeholder:** Who the key players are, their priorities, communication preferences, and relationship histories.
- **Decision:** Why the customer chose this solution—both the stated and unstated motivations.
- **Emotional:** Anxieties, expectations, and past experiences that shape the customer's approach.
- **Political:** Internal dynamics, potential resistance points, and champion relationships.
- **Trust:** Specific promises, assurances, or demonstrations that built confidence during the sales process.

For the cybersecurity implementation at GlobalDefense, this might include knowledge that the CISO had previously experienced a failed security implementation with a different vendor, that the IT Director initially opposed the zero-trust approach but was convinced through specific technical demonstrations, or that the board was particularly concerned about compliance with new government regulations taking effect during the implementation timeline.

Layer 2: Technical Transfer (Executional/Architectural)

The Technical Transfer layer focuses on the solution specifications, requirements, and implementation details. It ensures that the implementation team understands:

- **Functional:** Detailed specifications of what was sold and how it should work.
- **Engineering:** Technical assumptions, inputs/outputs, tolerances, and constraints.
- **Architecture:** System maps, workflow diagrams, integration points, and data flows.
- **Edge:** Known limitations, potential failure points, and technical risks.
- **Change:** Scope adjustments, requirement evolutions, and alternative approaches considered.
- **Implementation:** Required resources, prerequisites, and critical path items.

For SecureShield's handoff, this might include detailed network architecture diagrams showing integration points with GlobalDefense's existing security infrastructure, specific compliance requirements for different types of data, performance expectations under various threat scenarios, and compatibility requirements with legacy systems that couldn't be immediately replaced.

The Integration Point: Where Trust Meets Execution

The power of the Dual-Layer model comes from the integration of these two layers. Technical implementations don't fail merely due to technical misalignments; they fail because of disconnects between expectations and delivery, communication breakdowns, and trust erosion.

By addressing both layers explicitly, the Apex 8 Handoff creates a foundation where implementation teams:

- Understand not just what to build, but why it matters to the customer.
- Recognize potential political challenges that could derail technical progress.
- Appreciate the promises and expectations established during the sales process.
- Identify potential trust risks alongside technical risks.

When this integration is successful, customers feel seen and supported, implementation teams move with confidence, and the selling organization delivers on its promise—building momentum for both results and relationship growth.

This integration transforms handoffs from mechanical processes into strategic transitions that preserve and build upon the foundation established through the entire Apex 8 methodology.

The AI-Enhanced Handoff Process

This dual-layer framework is powerful on its own—but when paired with AI, it scales effortlessly, bringing new precision to what used to be inconsistent and informal.

In the AI era, handoffs become dramatically more comprehensive and effective through intelligent augmentation. The consciously competent seller leverages AI in several key ways:

1. Comprehensive Knowledge Synthesis

AI transforms the vast information accumulated throughout the Apex 8 process into structured, actionable handoff documentation:

- **Conversation:** AI analysis of all customer interactions (discovery sessions, validation discussions, presentations) to extract key insights, commitments, and concerns.
- **Document:** Automatic compilation and summarization of all project documents, proposals, and specifications.
- **Pattern:** Identification of recurring themes, priorities, and potential risks across the entire sales process.
- **Stakeholder:** Creation of detailed stakeholder profiles based on interaction patterns and stated preferences.

For the cybersecurity handoff, AI can synthesize numerous technical conversations to identify specific security concerns that were not explicitly documented but frequently appeared in discussions, or to recognize patterns in how different stakeholders described their priorities.

2. Dual-Layer Package Creation

AI automatically generates customized handoff documentation tailored to various roles and needs:

- **Trust:** Stakeholder profiles, relationship maps, decision journey documentation, and communication guides.
- **Technical:** Requirement specifications, architecture diagrams, integration maps, and implementation checklists.
- **Customized:** Tailor the handoff for each person—provide engineers with diagrams, project managers with timelines, and executives with strategic highlights.
- **Multi-format:** Documentation adapted for various consumption methods (presentations, reference guides, quick-access summaries).

The AI not only compiles information; it organizes it according to implementation priority and risk level, ensuring that critical information is prominently highlighted.

3. Continuity Mechanisms

AI creates active systems that maintain knowledge continuity throughout implementation:

- **Living:** Handoff materials that update automatically as implementation progresses.
- **Context:** Systems that maintain the history of decisions and alternatives considered.
- **Question:** Predictive frameworks that identify likely implementation questions and prepare answers.
- **Knowledge:** Natural language interfaces that allow implementation teams to query the collective project intelligence.

When a change request emerged halfway through GlobalDefense's deployment, the implementation team used living documentation to trace why a specific integration path was chosen—avoiding costly rework and reinforcing trust.

This transforms handoffs from one-time transfers into continuous knowledge flows that support implementation teams throughout the delivery process.

4. Risk Identification and Mitigation

AI analyzes the entire sales process to identify potential implementation risks:

- **Promise:** Verification that all commitments made during sales are reflected in implementation plans.
- **Expectation:** Identification of potential gaps between customer expectations and delivery capabilities.
- **Implementation:** Comparison with similar past projects to identify common challenges.
- **Resource:** Assessment of whether allocated resources align with project requirements.

For SecureShield, this might include detecting that certain security features demonstrated during sales would require additional configuration not explicitly included in the statement of work, or recognizing that the timeline expectations set with GlobalDefense's executive team might be challenging given the company's historical deployment patterns.

AI enables us to mitigate risk more precisely than ever, but the most reliable safeguard is a culture where handoffs—win or lose—are treated as essential and strategic.

The Universal Handoff Requirement

While we've focused primarily on handoffs for won opportunities, the Apex 8 methodology recognizes that every opportunity—whether won or lost—requires an appropriate handoff to relevant stakeholders.

Handoff for Won Opportunities

When you win, the primary handoff is to the implementation team, focusing on:

- Transferring customer relationships and context.
- Communicating solution specifications and requirements.
- Identifying potential implementation risks and challenges.
- Establishing continuity mechanisms for ongoing support.

See 'Continuity Mechanisms' for detailed strategies to reinforce these practices with structure and rhythm.

This ensures that customer value is realized through successful implementation.

Handoff for Lost Opportunities

Handoffs for lost deals aren't about revisiting failure—they're about harvesting competitive insights, customer feedback, and lessons that fuel smarter future wins.

When you lose, handoffs take different forms but remain equally important:

- **Product**: Communicating competitive gaps, feature requirements, or pricing issues that contributed to the loss.

- **Marketing**: Providing insights about messaging effectiveness, competitive positioning, or market perceptions.
- **Strategy**: Delivering intelligence about market trends, competitive tactics, or customer priorities to leadership.
- **Sales**: Transferring learnings about effective (or ineffective) sales approaches to improve future opportunities.

These handoffs transform losses into organizational learning and improvement opportunities.

Internal vs. External Handoffs

The Apex 8 methodology distinguishes between internal handoffs (to other teams within your organization) and external handoffs (to customers or partners):

Internal Handoffs focus

- Complete transfer of customer and solution knowledge.
- Identification of risks and special considerations.
- Creation of accountability for delivery.
- Preservation of customer trust through implementation.

External Handoffs focus

- Setting appropriate expectations for next steps.
- Introducing implementation team members and roles.
- Establishing communication protocols for the implementation phase.
- Creating confidence in the transition process.

Both types require deliberate attention and structured approaches to be effective.

From Handoff to New Discovery

The Handoff step doesn't merely conclude the Apex 8 methodology—it connects back to Discovery, creating a continuous cycle of value creation.

The Implementation Opportunity Loop

Successful implementations create opportunities for expansion, enhancement, and evolution:

- **Expansion**: Identifying additional areas where the solution could provide value.
- **Performance**: Understanding how the implemented solution is performing relative to expectations.
- **Evolution**: Exploring how changing customer needs might require solution adaptation.

The consciously competent seller recognizes that implementation success creates the foundation for new opportunities and begins the Discovery process again from a position of established trust and demonstrated value.

The Learning Loop

Even when no new opportunity immediately follows, implementation becomes a classroom—refining the product, sharpening the approach, and tuning future targeting. This is where strategy is shaped by reality.

In contrast, another project stalled when implementation challenges were never fed back to the product team—resulting in repeated friction for future buyers facing the same conditions.

These loops ensure that the organization continuously improves based on implementation experiences, creating a competitive advantage through accumulated wisdom.

For SecureShield, the GlobalDefense implementation might reveal that certain security features were particularly valuable to defense contractors, informing both product development priorities and sales approaches for similar prospects. It might also identify integration challenges that could be addressed earlier in the sales process for future opportunities.

Completing the Apex 8 Journey

As we reach the conclusion of the Handoff step, we complete the Apex 8 methodology journey. Let's briefly review the full methodology to understand how each step contributes to the complete cycle of value creation:

1. **Discover**: Uncovering the customer's current reality, challenges, and vision through deep exploration and active listening.
2. **Prototype**: Creating collaborative solution concepts that address the customer's specific needs and engage them as co-creators.
3. **Validate**: Confirming solution fit and value through evidence, testing, and verification in the customer's specific context.
4. **Rehearse**: Preparing thoroughly for an effective solution presentation, ensuring all aspects are refined and ready.
5. **Present**: Delivering a compelling, tailored narrative that connects technical capabilities to business outcomes.

6. **Finalize**: Moving from consideration to commitment by leveraging trust to help buyers confidently move past fear.
7. **Debrief**: Extracting maximum learning from each opportunity through structured reflection and analysis.
8. **Handoff**: Ensuring successful value realization by transferring both trust and technical requirements to implementation teams.

This comprehensive methodology transforms technical selling from a transactional process into a value creation cycle that benefits customers, implementation teams, and the selling organization itself.

You've not only learned a method—you've stepped into a new identity as a technical seller who leads with insight, delivers with integrity, and learns with purpose.

The consciously competent technical seller who masters all eight steps creates a virtuous cycle where:

- Customers receive solutions that truly address their needs.
- Implementation teams have the knowledge and context to deliver effectively.
- The selling organization continuously learns and improves.
- Trust builds across the entire cycle, creating foundations for ongoing relationships.

Applying Handoff Principles in Practice

Let's conclude with practical applications of these handoff principles in your next opportunity:

Dual-Layer Handoff Model

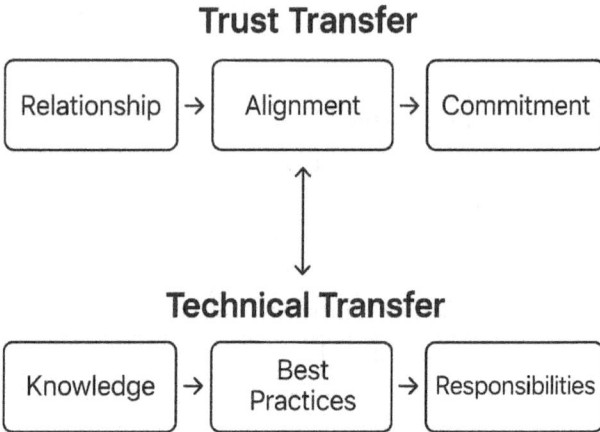

Trust Transfer

Relationship	→	Alignment	→	Commitment

↑
↓

Technical Transfer

Knowledge	→	Best Practices	→	Responsibilities

For your next won opportunity:

1. Create distinct Trust Transfer and Technical Transfer documentation.
2. Use AI to synthesize insights from the entire sales process.
3. Customize handoff materials for different implementation roles.
4. Include both the "what" and the "why" of solution decisions.
5. Explicitly identify potential implementation risks and challenges.

Leverage AI Effectively

Use AI to synthesize documents, identify risks, and generate tailored handoff kits—then review and refine them through human judgment.

To enhance your handoffs with AI:

1. Feed all customer interactions, documents, and decisions into your AI system.
2. Create standardized handoff templates that AI can populate automatically.
3. Use AI to identify patterns and risks across the sales process.
4. Develop systems for ongoing knowledge retrieval during implementation.
5. Create accountability tracking for handoff commitments.

Establish Continuity Mechanisms

To ensure smooth transitions:

1. Kick off with a joint session that sets the tone for trust.
2. Hand off relationships with care—people over process.
3. Co-create your communication rhythms.
4. Flag issues early, escalating with clarity.
5. Maintain a friendly connection between sales and delivery.

Build Organizational Discipline

To create a culture of effective handoffs:

1. Make handoffs a formal, required step for all opportunities (won and lost).
2. Measure implementation success and connect it to sales performance metrics.

3. Create feedback loops from implementation teams to sales teams.
4. Recognize and reward exemplary handoffs that lead to implementation success.
5. Continuously refine your handoff approach based on implementation outcomes.

Remember, the handoff isn't the end of the process—it's the bridge that connects sales promises to delivery reality. As sales leader Jim Dickie puts it: "Revenue isn't real until the customer experiences the value they were promised."

Make this your standard—not just for process, but for leadership. When you own the handoff, you elevate the entire delivery experience.

The Apex 8 methodology recognizes this fundamental truth and transforms handoffs from administrative afterthoughts into strategic imperatives that ensure customer success, organizational learning, and sustained competitive advantage.

The New Technical Seller

As we conclude our exploration of the Apex 8 methodology, let's reflect on what it means to be a technical seller in the AI era:

The new technical seller isn't merely a presenter of capabilities or a negotiator of terms. They are:

- A **trusted** who deeply understands customer realities.
- A **solution** who co-creates value with customers.
- A **value** who proves outcomes before commitment.
- A **preparation** who rehearses for maximum impact.

- A **compelling** who connects technology to outcomes.
- A **confidence** who helps buyers move past fear.
- A **reflective** who continuously learns and improves.
- A **value transfer** who ensures implementation success.
- A **continuous** who turns every engagement into feedback and fuel.

Each of these roles isn't abstract—they embody the eight steps: you discover as an advisor, prototype as an architect, validate as a builder of belief, and so on.

By mastering the Apex 8 methodology, you become this new type of technical seller—one who thrives in the AI era by combining technological augmentation with the uniquely human capabilities of wisdom, judgment, empathy, and creative problem-solving.

The future belongs to those who can leverage AI to handle the routine and analytical aspects of selling while elevating their focus to the strategic and relational elements that create lasting value. The Apex 8 methodology provides the framework for this transformation, giving you the conscious competence to excel in a rapidly evolving landscape.

As you apply these principles in your own selling context, remember that the methodology is not a rigid formula but an adaptive framework. The specific techniques and tools will evolve, but the fundamental principles—understanding before solution, collaboration before presentation, validation before commitment, reflection after completion—will remain powerful guides for success.

The journey of mastery is ongoing. Each cycle through the Apex 8 methodology builds your expertise, refines your approach, and strengthens your ability to create value for customers in increasingly complex technical environments.

The future of technical selling has arrived. With the Apex 8 methodology as your guide, you're ready to meet it.

CHAPTER 10

THE JOURNEY BEYOND: YOUR PATH TO MASTERY

The Journey From Knowledge to Practice

As we conclude our exploration of the Apex 8 methodology, the question becomes: where do you go from here? How do you transform the knowledge you've gained into a lived practice that revolutionizes your approach to technical selling?

Throughout this book, we've deliberately structured the content to achieve two objectives: building your conscious competence about the methodology itself while creating the foundation for those moments of unconscious excellence—the "in the zone" states where your expertise flows naturally without conscious effort.

The Mastery Cycle: Conscious to Unconscious Competence

As you move from reading about the Apex 8 methodology to implementing it, you'll engage in a continuous cycle between conscious and unconscious competence—a fundamental rhythm we explored in Chapters 5 and 6.

This cycling isn't random or haphazard. The ability to shift between structured thinking and intuitive action isn't just elegant—it's essential in the real world, where every conversation, stakeholder, and context demands a slightly different version of you. Different phases of technical selling demand different competence states:

Discovery begins with conscious competence as you deliberately apply questioning frameworks, but may shift to unconscious competence as you develop intuitive listening skills that detect subtle signals.

Prototyping requires conscious competence for structured development, balanced with unconscious competence for creative flow when generating innovative solution approaches.

Validation benefits from rigorous conscious competence to ensure thoroughness, with moments of unconscious intuition when identifying unstated concerns.

Rehearsal/Simulation builds the confidence foundation that allows unconscious competence to emerge during actual presentations. As we explored in Chapter 5, this simulation isn't about memorization but about creating the conditions that allow you to "get in the zone" when it matters most.

Presentation/Emulation involves a fluid dance between conscious structure and unconscious flow. As we discussed in Chapter 6, the most compelling presentations happen when you can shift seamlessly between deliberate technique and authentic, in-the-zone moments where your expertise emerges naturally.

Finalization requires conscious competence for strategic planning, coupled with unconscious competence for reading subtle buyer signals and responding authentically to concerns.

Debrief demands a return to conscious competence for systematic analysis, extracting maximum learning from each opportunity.

Handoff needs conscious competence to ensure complete knowledge transfer, though relationships may benefit from the authentic connection that comes from unconscious competence.

This dynamic shifting between states is the hallmark of mastery in technical selling. The novice rigidly applies techniques without flexibility. The master knows when to deliberately apply frameworks and when to trust their expertise to flow naturally.

The Transformative Meta-Moment

Throughout this book, we've created opportunities for meta-moments—those higher-level recognitions where you see beyond immediate techniques to the transformative patterns and possibilities of the methodology as a whole.

Some readers will have already experienced these meta-moments—recognizing not just the specific techniques and approaches but the profound shift in perspective they represent. You may have seen how Apex 8 isn't just a sequence of steps but a fundamentally different paradigm for creating value through technical selling.

Others may still be processing the methodology on a more practical level—understanding the steps and techniques but not yet experiencing that transcendent recognition. If you're in this group, don't be discouraged. Meta-moments can't be forced; they emerge through engagement, practice, and reflection.

The beauty of the methodology is that it works even before you've fully grasped its meta-implications. By following the structured approach—moving systematically through Discovery, Prototyping, Validation, Rehearsal, Presentation, Finalization, Debrief, and Handoff—you create value regardless of whether you've experienced the higher-level recognition.

But when that meta-moment comes—when you suddenly see how all eight steps create a seamless cycle of value creation that transforms your relationship with selling itself—your application of the methodology reaches a new level of power and integration.

First Steps: Beginning Your Practice

As you close this book and begin implementing the Apex 8 methodology, the most powerful first step is deceptively simple: return to Chapter 2 and focus on Discovery.

Why Discovery? Because listening is the foundation of everything that follows. Mastering technical selling starts with mastering attention—the ability to be fully present, genuinely curious, and deeply attentive to what others communicate.

The beauty of beginning with Discovery is that you can start immediately. Your very next conversation is an opportunity to practice. Every interaction becomes a chance to enhance your listening skills, ask better questions, and uncover deeper insights.

Here's a liberating truth: it's okay to make mistakes. It's acceptable if your first attempts at structured Discovery feel awkward, or if you miss important signals. It's fine if your initial Prototyping efforts don't perfectly address customer needs. It's alright if your early Validation approaches leave gaps.

These aren't failures; they are essential learning experiences that fuel growth. Every master began as a novice, and every expert started with mistakes. The difference lies in how they responded—not with discouragement but with curiosity, not with retreat but with refinement.

While your first step begins with human presence and attention, your path forward can be accelerated by a powerful partner: artificial intelligence.

AI: Your Accelerating Partner

As you embark on this journey to master the Apex 8 methodology, artificial intelligence becomes your partner in accelerating the path to mastery.

The true power of AI in technical selling lies not in automation but in acceleration—compressing the learning cycle that traditionally took years or decades into months or even weeks. AI helps you:

- **Refine:** Analyzing conversation patterns to identify missed signals and opportunities.
- **Enhance:** Rapidly generating customized solution approaches for different scenarios.
- **Strengthen:** Identifying patterns and potential gaps in evidence.
- **Elevate:** Creating sophisticated simulation environments to build confidence.

- **Improve:** Providing real-time insights and adaptation suggestions.
- **Optimize:** Detecting subtle commitment signals and concern patterns.
- **Deepen:** Extracting patterns and insights across multiple opportunities.
- **Streamline:** Creating comprehensive knowledge packages for implementation teams.

As we discussed earlier, AI shortens the learning curve—turning feedback into forward motion and helping you move faster through the mastery cycle.

But remember—AI is an accelerator, not a replacement. It enhances but doesn't substitute for your uniquely human capabilities: your empathy, judgment, creativity, ethical reasoning, and relationship building. These remain the differentiators that no algorithm can replicate.

The most effective technical sellers in the AI era are neither AI-resistant traditionalists nor AI-dependent technicians. They are AI-augmented experts who leverage technology to enhance their human capabilities while developing the wisdom to know when to trust their unconscious expertise to guide them.

The Transformation: From Theory to Practice

Let me share the story of Alex Chen, a 28-year-old technical sales representative at NeuraTech, a company selling AI-enhanced data analytics platforms to enterprise clients.

Alex had always been technically proficient. With a computer science degree and two years of experience as a solutions engineer before moving into sales, he understood the product inside and out. However, six months into his sales role, his results were mediocre at best. He could explain technical specifications flawlessly but struggled to connect with decision-makers and close deals.

His manager suggested he attend a workshop on the Apex 8 methodology, hoping it might help him bridge the gap between technical knowledge and sales effectiveness. Alex was skeptical—he had tried generic sales techniques before, and they felt inauthentic when applied to complex technical solutions.

Nevertheless, he attended the workshop, took detailed notes, and decided to give the approach a chance with his most challenging prospect: GlobalFinance, a financial services firm that had been evaluating analytics platforms for months without making a decision.

Discovery

Instead of beginning his next call with GlobalFinance by discussing NeuraTech's latest features, Alex prepared a structured discovery approach. He used his AI sales assistant to analyze previous interactions, identifying potential gaps in his understanding of their situation.

On the call, he opened with: "Before we dive into any specifics today, I'd like to revisit your core challenges to ensure I fully understand what you're trying to accomplish."

For the next thirty minutes, he mostly listened, asking targeted questions about their current analytics processes, challenges with data integration, and specific outcomes they hoped to achieve. His

AI assistant analyzed the conversation in real-time, suggesting follow-up questions when it detected uncertainty or inconsistency in the responses.

What Alex discovered surprised him. While he had been focusing on processing speed and algorithm sophistication—technical aspects where NeuraTech excelled—GlobalFinance's primary concern was actually user adoption. Previous analytics investments had delivered technical capabilities but failed because front-line employees found them too complex to use effectively.

Prototype

Instead of moving straight to a standard demonstration, Alex worked with GlobalFinance to prototype a solution specifically addressing user adoption. Together, they sketched user interface workflows, created sample dashboards tailored to different user roles, and developed a phased implementation approach emphasizing training and adoption metrics.

Using his AI-powered prototyping tool, Alex was able to transform rough concepts into visual mock-ups during the meeting itself. When the CIO mentioned a specific reporting challenge, Alex could immediately create a prototype dashboard showing how NeuraTech would address it.

"This is the first time a vendor has actually shown us something that looks like what we need, not just what they want to sell us," the CIO commented.

Validation Through

Before presenting a formal proposal, Alex focused on validating the approach. He arranged for GlobalFinance's team to speak directly with similar clients who had faced adoption challenges. He provided access to a limited trial environment where their team could test the user experience themselves. He worked with their data science team to validate integration approaches using sample data sets.

Each validation step built confidence while also refining the solution. When concerns emerged about mobile access, Alex adjusted the approach to prioritize mobile-friendly interfaces for senior executives.

Rehearsal for

Before his final presentation to GlobalFinance's executive committee, Alex conducted thorough simulations. He practiced with his own team, incorporating their feedback. He used his AI presentation coach to analyze his delivery, identifying moments where he reverted to technical jargon or lost his connection to business outcomes.

Most importantly, these simulations established the confidence foundation that would allow him to enter "the zone" during crucial moments in the actual presentation. Rather than trying to memorize responses to every possible question, Alex developed the confidence to trust his expertise to flow naturally when it mattered most.

Presentation That

When presentation day arrived, Alex didn't deliver a standard capabilities overview. Instead, he told the story of GlobalFinance's analytics journey—from where they were to where they wanted to be—with NeuraTech as the enabling partner in that transformation.

Throughout the presentation, Alex moved fluidly between conscious competence and unconscious flow. During structured segments explaining the solution architecture, he deliberately applied the frameworks he had practiced. But when the COO interrupted with concerns about implementation timelines, Alex smoothly transitioned into "the zone"—responding with authentic expertise that addressed the concern while maintaining a natural connection.

Finalization Without

As the presentation concluded and discussion turned to next steps, Alex recognized hesitation from the CFO—not about the solution itself, but about the implementation timeline. Instead of pushing for immediate commitment, Alex acknowledged the concern directly:

"I sense you have some reservations about the timeline we've outlined. Is it the pace of implementation or something else that concerns you?"

The CFO explained that a previous failed implementation had created skepticism about aggressive timelines. Rather than dismissing this concern or applying pressure, Alex suggested a modified approach with a phased rollout, smaller initial investment, and clear success metrics before expanding.

This addressed the fear without diminishing the solution, creating a path to commitment that felt safe and realistic.

Debrief for

When GlobalFinance signed the agreement—larger in scope than any of Alex's previous deals—his first response wasn't celebration but reflection. He conducted a thorough debrief with his team and his AI assistant, analyzing what worked, what could be improved, and what insights could apply to other opportunities.

He identified that collaborative prototyping had been particularly effective and that directly addressing adoption concerns differentiated NeuraTech from competitors who focused primarily on technical capabilities.

Handoff for

Alex's final step was ensuring a smooth handoff to the implementation team. He created a comprehensive dual-layer handoff package, transferring both the technical specifications and the relationship context to the team that would deliver the solution.

He arranged a joint meeting with GlobalFinance stakeholders and the implementation team, ensuring continuity of understanding and relationships. His AI assistant generated a detailed knowledge graph of all interactions, decisions, and commitments, ensuring nothing would be lost in the transition.

The Ripple

Six months later, GlobalFinance's implementation was succeeding beyond expectations. User adoption reached 78% in the first phase—unprecedented for their analytics initiatives. The initial success led to expansion into additional departments, doubling the original contract value.

However, the transformation wasn't limited to this single opportunity. Alex's approach to other prospects evolved dramatically. His pipeline conversion rate tripled, and his average deal size increased by 60%. While the average deal size in his segment hovered around $140K, Alex's post-Apex average exceeded $220K—proof that his new approach was more than just personal growth; it was business transformation. He moved from the middle of the sales rankings to consistently leading the team.

More significantly, his relationship with his work transformed. Technical selling was no longer about pushing features or navigating uncomfortable negotiations. It became a collaborative partnership with clients to solve meaningful problems—more fulfilling professionally and more effective commercially.

When his manager asked what had changed, Alex's answer was simple: "I stopped selling products and started creating value through a methodology that made sense for technical solutions. And I learned when to consciously apply techniques and when to trust myself to be in the zone—letting my authentic expertise flow naturally."

Zooming Out: From Selling to Commercialization

As you master the Apex 8 methodology, it's worth taking a step back to understand where technical selling fits within the broader landscape of value creation. What we've explored in this book—the eight-step methodology for technical selling—is a crucial component of a larger process: commercialization.

While selling focuses on the transaction—converting leads into paying customers for solutions that are ready for market—commercialization encompasses a more strategic and systemic approach to bringing value to market.

Selling vs. Commercialization: The Essential Selling:

- Transactional in nature
- Focused on converting leads to customers
- Often begins after the product is built
- Primarily concerned with immediate revenue
- What we've explored throughout this book with Apex 8

Commercialization:

- Strategic and systemic
- Focused on bringing an idea to market and making it sustainable
- Begins before the product is built—includes product-market fit, positioning, and go-to-market strategy
- Oriented toward recurring revenue
- The broader context within which selling operates

By commercialization, we don't just mean monetization; we mean shaping what gets built, how it's priced, positioned, supported, and scaled.

As you grow in your mastery of Apex 8, you may find yourself naturally expanding your perspective from the transaction-focused approach of selling to the more strategic view of commercialization. In many ways, Debrief and Handoff lay the groundwork for this evolution—translating frontline experience into strategic insight that informs not just deals, but the direction of products and markets. This evolution often occurs as technical sellers develop deeper expertise about markets, customer needs, and solution possibilities.

The AI era has accelerated this evolution by enabling sellers to process and synthesize vast amounts of market intelligence, identify patterns across customer engagements, and contribute strategic insights to product development and positioning. What once might have taken a career to develop can now emerge through AI-accelerated learning in a fraction of the time.

This broader perspective introduces the concept of **Value** — where commercialization strategy is blended with technical architecture to grow, expand, or accelerate the success of an organization. Value Architecture creates the structure that bridges human expertise and AI capabilities, something that was difficult to achieve before the AI era arrived. It represents the next evolution beyond transaction-focused selling: designing integrated systems of technology, expertise, and commercial strategy that deliver repeatable, scalable value.

As you continue your journey beyond this book, consider how your developing expertise might contribute not just to individual sales but to the broader commercialization of your solutions and potentially to the design of value architecture that creates sustainable success in the marketplace.

The Apex 8 Community

Your journey with the Apex 8 methodology doesn't have to be solitary. A growing community of practitioners across industries is implementing, refining, and extending the approach—sharing insights, overcoming challenges, and celebrating successes together.

Consider connecting with this community through:

- **Online:** Spaces where practitioners discuss specific challenges and innovations
- **Local:** In-person gatherings for relationship building and experience sharing
- **Training:** Structured learning environments for deepening your mastery
- **Certification:** Formal recognition of your expertise in the methodology

These connections not only accelerate your learning but create a support system for the inevitable challenges of implementation. They provide external perspectives when you get stuck and celebrations when you achieve breakthrough results.

The methodology continues to evolve through the collective wisdom of this community. Your experiences and insights—both successes and setbacks—contribute to this evolution, refining the approach for yourself and others. In those moments when doubt creeps in or momentum wanes, having a group that speaks your language and shares your mission can be the difference between staying stuck and taking your next leap.

That sense of shared practice—of growing alongside others who are walking the same path—is what gives your solo journey fuel and forward motion.

Your Journey Begins

As you close this book and begin your own journey with Apex 8, remember these core principles:

1. **Anchor:** Because presence precedes progress

2. **Master**: So your expertise flows with agility and power
3. **Create**: To enter the zone when it matters most
4. **Leverage**: To accelerate without losing your humanity
5. **Reframe**: As raw material for mastery
6. **Connect**: Because learning is faster (and more fun) together

The transformation available to you isn't just professional; it's personal. Beyond improving metrics like win rates and deal sizes, the Apex 8 methodology can fundamentally change your relationship with selling itself. Technical selling becomes not a necessary evil or a challenging obligation but a fulfilling practice of creating genuine value through authentic connection.

Like Alex, you're stepping into your own story now—one that's shaped not just by knowledge, but by how you show up, adapt, and grow with intention.

You've been introduced to the methodology throughout this book. Now it's yours to implement, adapt, and master. The journey from understanding to application begins with your very next conversation.

This isn't the end—it's just the beginning. And what a remarkable beginning it can be.

THE VALUE ARCHITECT'S PERSPECTIVE

"Selling is about the transaction. Commercialization is about unlocking scalable value. I don't just sell solutions to grow companies—I commercialize them. That means building the systems, messaging, and market fit that make AI-enabled growth repeatable, profitable, and real."

— Daren Fields, April 2025

As we conclude our exploration of the Apex 8 methodology, I'd like to share a perspective on what might lie beyond mastery of technical selling—a vision of how your expertise might evolve as you continue your professional journey.

Most Value Architects don't start with that title. They grow into it through questions they couldn't stop asking, patterns they started seeing, and value they couldn't ignore.

The Emergence of the Value Architect

In the AI era, a new role is emerging at the intersection of technology, sales, and strategic business development: the Value Architect. This role transcends traditional boundaries between technical expertise, sales prowess, and business strategy to create comprehensive frameworks for sustainable value creation.

Value is a relatively new term that represents a powerful concept: the deliberate blending of commercialization strategy with technical architecture to grow, expand, or accelerate the success of an organization. While related concepts like value engineering have existed for some time, they don't fully capture the technology integration aspect that is central to value architecture.

Value Architecture creates a structured bridge between:

- The technical capabilities of products and solutions
- The human expertise and processes that deliver them
- The commercial strategy that brings them to market
- The scalable systems that make success repeatable

The Value Architect designs how these elements work together, creating a blueprint for value creation that integrates both technological and human components into a cohesive system that delivers consistent, scalable results.

Why Now? The AI Catalyst

The emergence of the Value Architect role has been catalyzed by artificial intelligence for two critical reasons:

First, AI has dramatically expanded the possibilities in technical solutions, creating opportunities for value creation that were previously inconceivable. These possibilities require someone who can envision, articulate, and orchestrate complex value systems that integrate both technical and human elements.

Second, the partnership between humans and AI has become a critical component of value architecture. Designing how AI and human capabilities work together to create maximum value for customers has become a specialized expertise that combines deep technical understanding with strategic business vision.

In the past, orchestrating both the technical and commercial layers of value was too complex for most individuals. However, AI now enables experts to see further, synthesize faster, and operate across boundaries that were once siloed. What was once nearly impossible—simultaneously mastering the technical complexities of solutions while orchestrating their commercial success—has become achievable through AI augmentation. The Value Architect leverages AI to extend their capabilities across both domains, creating integrated approaches to value creation.

From Apex 8 Mastery to Value Architecture

The path from mastery of the Apex 8 methodology to Value Architecture is a natural evolution for those with the vision and ambition to expand their impact.

As you become increasingly proficient in technical selling through Apex 8, you develop:

- A deep understanding of how technical capabilities translate to business value
- Insight into market patterns and customer needs across industries or segments
- Expertise in articulating complex value propositions effectively
- Experience in designing solution approaches that integrate technical and human elements
- Wisdom about successful implementation and value realization

These capabilities form the foundation for transitioning from transaction-focused selling to strategic value architecture.

Each step in Apex 8 develops one dimension of this capability—Discovery shapes insight, Prototype builds structure, Validate sharpens clarity, and Handoff cements credibility. Together, they form the muscle memory of value creation.

The transition often begins with contributing insights beyond individual sales opportunities—identifying patterns across customer needs, suggesting refinements to product capabilities or positioning, or developing repeatable approaches to value demonstration.

Over time, this expanded perspective may evolve into a formal or informal role as a Value Architect—someone who shapes not just how solutions are sold but how they are conceived, developed, positioned, and commercialized.

So, what does it take to evolve into a Value Architect? Here's how the disciplines begin to align.

The Value Architect's Toolkit

Value Architects blend tools and approaches from multiple disciplines:

From Technical architecture

- Systems thinking and integration design
- Scalability and performance optimization
- Technical dependency mapping
- Platform and ecosystem design
- Designing workflows

From Sales and Marketing

- Value proposition development
- Narrative and messaging frameworks
- Buyer journey mapping
- Trust-building methodologies
- Crafting lens

From Business strategy

- Market opportunity analysis
- Competitive positioning
- Business model innovation
- Strategic partnership development
- Identifying clients

From implementation

- Success pattern recognition
- Risk mitigation frameworks
- Adoption acceleration approaches
- Value measurement systems
- Developing process

The power comes not from mastering each discipline individually but from integrating them into cohesive frameworks for value creation and realization.

Why Value Architecture Matters

As explored in the toolkit above, advantage now comes from integration—where systems of insight, delivery, and design align around value.

Organizations that master value architecture can:

- Create solutions that deliver value more consistently and predictably
- Accelerate market adoption through clearer articulation of differentiated value
- Build more sustainable competitive advantages that transcend feature comparisons
- Develop repeatable, scalable approaches to value creation across markets
- Create virtuous cycles of innovation based on a deeper understanding of value patterns

When value becomes systemic rather than sporadic, you don't just win deals—you create momentum, loyalty, and long-term impact.

For individual professionals, developing value architecture expertise opens paths to expanded impact and influence—roles that shape not just individual transactions but the strategic direction of products, services, and organizations.

The Future of Value Creation

You've already seen how systems—not just features—are the new frontier. In the chapters ahead, this truth will only deepen as AI reshapes every layer of how we define and deliver value. The greatest opportunities will belong to those who can envision and orchestrate integrated systems of value that seamlessly blend technical capabilities, human expertise, and business models.

The Apex 8 methodology you've explored in this book provides a solid foundation for this journey—a systematic approach to technical selling that can evolve into broader expertise in value creation and commercialization.

Where will your journey take you? From mastery of technical selling to leadership in value architecture? From transaction expertise to commercialization strategy? From selling products to architecting value systems?

Whatever path you choose, remember that the principles you've explored in Apex 8—conscious competence, continuous discovery, collaborative development, systematic validation, thorough preparation, compelling presentation, fear resolution, and learning through reflection—remain powerful guides for creating exceptional value in an ever-evolving landscape.

The future belongs not just to those who can sell technology but to those who can architect value. And that journey begins with mastery of the fundamentals you've explored in these pages.

ACKNOWLEDGEMENTS

To my parents, who fostered my entrepreneurial spirit even when it led me down paths they themselves would not have traveled. Thank you for never doubting my ability to make an impact and for encouraging me to find my own way.

To my motorsports and motorcycle brethren who taught me discipline and excellence: Jeff & Kirk Hoeppner, Peter Carroll, Brad Fairbanks, Dean Mizdal, Pat Burman, Paul Moen, Michael Henao, Terry Statum, Marcus Falley, Todd Brubaker, Keith Kanzel, Frank Belio, Keith Code, and David Sadowski, and many other two and four-wheel legends that I was lucky enough to meet, even if just for a minute. You all taught me the power of tenacity and you always encouraged me to be an innovator, to take chances and to never hold back. You taught me that there was a precision to everything that needed to be done in order to be successful, otherwise it just wasn't going to go the way you wanted it to go. You all taught me that being fearless in this world is really the only way to live, and those lessons I learned on the track, around the track, and even off the track have translated into success in every aspect of my life.

Deep gratitude to my sales mentors who have profoundly influenced my approach to ethical sales and leadership: Mark Johnson, Kevin Durkin, Ed Allen, Robert Miller, Lior Weinstein, Vince Tortola, Joe Andrulis, and Fran Tarkenton. Your guidance and wisdom have been invaluable, shaping not just this book, but my entire professional philosophy.

To the technical people who taught me how to be a value architect: Terrence Callahan, Jan Baan, Charles Phillips, David Phillips, Scott Hine, Adam Combs, Paul Denmark, Grace Velker, Dan Hinkle, Michael Janes, Eric Bragg, Simon Eyre, Don Leonardo, David Pezzino, Michael Chesin, Marissa Brassfield, Alex White, Chris Griffith, Mario Gulang, Kyle Becker, and Mark Siner. Your expertise and willingness to share knowledge were invaluable in assisting me to close some of the largest deals ever secured in the software and technology industries. Your guidance transformed how I approached these multimillion dollar projects and created lasting value for all of our clients.

I am equally grateful to those who challenged my ideas and pushed me to defend my positions with greater clarity and conviction. Your skepticism strengthened my resolve and refined my thinking in ways that agreement never could.

To the altruistic and insightful authors whose books have shaped my thinking, and to the great salespeople who came before me—thank you for teaching me, a complete stranger, how to make an amazing living putting great products in the hands of so many good people. Your wisdom, shared through pages and stories, has been an invaluable gift.

This book stands as a testament to the collective wisdom of all these remarkable individuals who have illuminated the path forward.

With sincere appreciation,
Daren "The Professor" Fields

ABOUT THE AUTHOR

Daren L. Fields is a distinguished author in the field of business revenue. With years of experience as a public speaker and master sales trainer, Daren has established himself as a leader in sales methodology and innovation. His writing seamlessly blends practical insights with strategic thinking, making him a sought-after voice in the industry.

In his career, Daren has excelled as a sales leader and trainer, imparting knowledge and skills to countless professionals. His achievements include leading successful workshops and developing innovative sales strategies that have transformed businesses.

Beyond his writing and speaking engagements, Daren is a technology innovator and attributes this unique skill to his years as a former motorsports competitor. His passion for technology and commercialization has driven his professional pursuits, earning him recognition as a commercialization officer. Currently, Daren continues to explore new avenues for growth and is dedicated to sharing his expertise with the world.